DNA:

the elephant
in the lab

---

the truth about

the origin of life

ORSON WEDGWOOD

Copyright © 2019 by Orson Wedgwood.

The right of Orson Wedgwood to be identified as the author of this work has been asserted.
All rights reserved.

No part of this book may be reproduced, stored in a retrieval system, or transmitted, in any
form, or by any means (electronic, mechanical, photocopying, recording or otherwise) without
the prior written permission of the author, except in cases of brief quotations embodied
in reviews or articles. It may not be edited, amended, lent, resold, hired out, distributed or
otherwise circulated, without the publisher's written permission.

Permission can be obtained from www.orsonw.com

Published by Orson Wedgwood.

ISBN: 978-0-9563725-8-1

Cover design & interior formatting:
Mark Thomas / Coverness.com (via Reedsy.com)

ACKNOWLEDGEMENTS

This book has undergone a number of revisions. I would like to thank my editors, Susan Hughes and Lee Ogelsby for their professionalism and excellent work; my Beta readers, Sarah Mackness, Floyd Running, Tom Dare and Joseph Tcherkezian for giving me valuable insights; and mostly my wife Kirsty, whose patience, editorial advice and belief in this project have kept me going.

eventim
+ +
+ +

eventim.co.uk

002286315100501030010000
616 006367217701

Please see www.eventimapollo.com for camera policy

Chambers Touring and Metropolis Music Presents

Flight Of The Conchords

Sing Flight of the Conchords Tour
Latecomers may not be admitted
Doors 7pm - U15s with an adult

Eventim Apollo
Tuesday 03 July 2018

Stalls
New Full Price Ticket

Row X   **Seat 33**
£65.00   FEE £9.25

006367217701

Flight Of The Conchords
Eventim Apollo
Tuesday 03 July 2018
Stalls   X 33
New Full Price Ticket
Ev: 616   65.00
Or: 002120767B   FEE 9.25
Cu: 0777718579
Ti: 006367217T
Kemp Caroline

002286315100501030010000

CTSeventim

1. This ticket is issued subject to the rules and regulations of the Venue and the Event Promoter, which can be found on the venue and/or event website
2. This ticket will be void if exchanged or refunded after purchase unless authorised by the Venue or Event Promoter
3. Please check your tickets carefully on receipt and contact us immediately if you have a query
4. Our General Terms and Conditions and Purchase Policy for this ticket can be found at www.eventim.co.uk/terms

eventim.co.uk

More from eventim.co.uk

Facebook
facebook.com/eventimuk

Eventim Newsletter
eventim.co.uk/newsletter

Twitter
twitter.com/eventim_uk

Ticket Alarm
eventim.co.uk/ticketalarm

Pinterest
pinterest/eventimuk

Eventim Ticketnews
ticketnews.eventim.co.uk

0030034185 UK   0030034185 UK

0030034185 UK   0030034185 UK

TABLE OF CONTENTS

"Then you will know the truth, and the truth will set you free."

- John 8:32

CONVICTIONS

Is this book for you?

> *"A man with a conviction is a hard man to change. Tell him you disagree and he turns away. Show him facts or figures and he questions your sources. Appeal to logic and he fails to see your point ... Suppose that he is presented with evidence, unequivocal and undeniable evidence, that his belief is wrong: what will happen? The individual will frequently emerge, not only unshaken, but even more convinced of the truth of his beliefs than ever before."*
>
> *- Leon Festinger, Henry Riecken, and Stanley Schacter:*
> *When Prophecy Fails, 1957*

This conclusion, which the authors came to after researching cult beliefs, has been confirmed over and over again by psychology research, most famously through a series of experiments on students in Stanford University in the 1970s[1]. Once a belief has become ingrained, and a worldview established around that belief, then facts alone will rarely change that mind.

If you are an atheist and reading this, then you are probably saying "Exactly! Christians just won't look at the facts and see the truth. It drives me potty!" Likewise, if you believe in God, you will say the same about atheists…both sets of believers would argue that facts support their beliefs and worldview, and no matter how eloquently you articulate these facts, and no matter how hard you try to appear objective, the person who holds the opposing worldview just won't budge. They are stubborn in their "ignorance."

This book has one aim: to try to establish the truth on a very specific, but controversial subject, about which many people have very strongly held views or convictions. We will be seeking, in a user-friendly way, to answer the question, "What does the sum of our scientific knowledge about the origin of life, and specifically, the origin of DNA, tell us?" Does the scientific evidence point toward a random natural process as the source? Does it point to some intelligent being "initiating" life? Or does it point to nothing? Is the evidence inconclusive? However, given what psychologists have concluded, it's fair to say that if you hold a very strong view on this subject, these scientific findings will either serve to reinforce this view, or you will dismiss the evidence.

I wouldn't want anyone to feel cheated, so I am going to be up-front and say that if you are a passionate atheist, then this book may be very challenging to read. You would have to be very open-minded and genuinely curious to read it and not feel uncomfortable. This sense of discomfort may intensify to a buttock-clenching level because a number of atheist or agnostic scientists agree with many of the conclusions about the science we will be discussing. They will, of course, be cited. So if you are an atheist, maybe your goal should be to simply understand that the science contradicting your position may

---

1   Biased assimilation and attitude polarization: The effects of prior theories on subsequently considered evidence; Lord, C. G., Ross, L., & Lepper, M. R. Journal of Personality and Social Psychology, 37(11), 2098-2109. 1979.

have at least a smidgen of validity, even if you don't agree with the ultimate conclusions.

To everyone else who is curious about this fascinating topic, then the goal is to leave you significantly more enlightened about one of the most important questions facing mankind: "Just where on (or off) Earth do we come from?" We will be going on a journey of discovery together, one that I traveled alone a number of years ago when I really questioned the foundations of my own worldview. We will repeat that process here, but with even more rigor.

From my years in scientific research and having studied this subject in great detail, it amazes me that anyone can see the world differently from the way that I do…but then it would, wouldn't it. I have a strong opinion, or conviction, so undoubtedly fall into the category of person that Festinger describes in the opening quote. Please, I beg you, bear that in mind as we travel along this road of discovery. Don't take what is said as the unchallenged truth or fact; go on the internet if you question what is written in these pages …everything cited is in the public domain. If you do, don't just accept the first article that agrees with your position; that is called confirmation bias. Read as much as possible, and by writers who you may disagree with. Be curious. Be open-minded, but don't be duped.

Having said that, to have any credibility when discussing a topic like this, not only do you have to be extremely knowledgeable on the subject you are discussing, but any claims that are made must be balanced and backed up by evidence. In other words, if there is evidence against the claims that are made, you must be prepared to discuss, or at least acknowledge this up-front and only dismiss it if there is sufficient evidence to the contrary. This is the approach that I have attempted to take, so if you do decide to cross-check facts I cite, I feel confident you will find I have looked under every stone. Much of the scientific minutia of this I have relegated to the appendix so that the main text is not an endless, turgid tome, and for those curious enough to enter a rabbit hole inside a red herring, then the appendix is the place to be.

So in answer to the opening question, "Who is this for?", it is for anyone, regardless of baseline scientific knowledge, who wants to understand precisely what DNA is, how it works, and whether existing science supports the belief that random causes or intelligence are the most likely explanation for its appearance on Earth.

The way that we will go on this journey is to walk through the evidence for and against random causes as the source of life on Earth, just as we would in a court of law. In fact, that brings me to why I wrote this in the first place...I believe a grave wrongdoing has been committed.

1.BACKGROUND

1.1 THE WRONGDOING

"I can't believe…that we are still debating and still questioning whether life was a divine intervention or whether it was coming out of a natural process let alone, oh my goodness, a random process."

I heard these words when I was in Canada giving talks on the very subject of the origin of life. The speech, which included this statement, was delivered by Julie Payette, Canada's Governor General, on November 1, 2017 to a science conference in Ottawa. The words by themselves, without the benefit of hearing her tone, could be read in two ways. Maybe she meant, "Oh my goodness, how could anyone still believe that life _was_ due to a random process?" or, "Oh my goodness how could anyone still believe that life _wasn't_ due to a random process?" When you hear her tone, dripping with theatrical condescension, it's perfectly obvious that she meant the latter, or more likely: "Anyone who believes that anything other than a random process initiated life is a knuckle-dragging halfwit who crawled out from beneath a rock, having missed every scientific discovery since the Enlightenment."

Now if this had been yet another rant by science writer Richard Dawkins[2] ruling out intelligent "input," it wouldn't have motivated me to write a book discussing whether there is evidence to suggest otherwise. But it wasn't Dawkins; it was the Queen's representative in Canada, a member of the ruling establishment, the global elite, implying there is no place in government or academia for questioning the belief that a random natural process generated life. I suspect the Queen would have been displeased when she heard that one of her servants spoke out of turn on a subject on which she herself has a very clear position, not least because she is the head of the Anglican Church, but also because she has stated on many occasions how much her Christian faith means to her. Moreover, judging by the controversy that this event caused in a normally polite and civil country, many Canadians also felt it was not her place to deride the beliefs of her fellow citizens.

But Payette's careless remark was far from the worst aspect of this little episode that I witnessed while in Canada. Justin Trudeau, Canada's Prime Minister, issued a statement supporting Payette which concluded:

"I applaud the firmness with which she stands in support of science and the truth."

The truth? Really?

Andrew Sheer, the leader of the Canadian opposition, pointed out that Payette had just offended many people of faith in Canada. While it was good to see him defend religious liberty, he didn't defend "the truth", possibly

---

2   Richard Dawkins is one of the leading lights of New Atheism and one of the "Four Horsemen of Atheism" which also included Christopher Hitchens, Daniel Dennett, and Sam Harris. Dawkins and other members of this group may have fallen out of favor with the establishment because of some comments that were made which perhaps didn't conform to the 'approved' position on Islam.

because, like so many people, he doesn't really know exactly what the truth on this subject is himself. It was at this point that I decided something more had to be done. Initially I went all "Alex Jones"[3] and produced a swivel-eyed YouTube video with an argument along the lines of Payette and Trudeau being the two-headed embodiment of Satan, stalking the Earth and devouring good Christian people. Then I calmed down, deleted the video before anyone could see it and thought about a more mature response. That is when I started to write this manuscript.

As you've probably guessed, I am a person of faith (full disclosure, I am a Christian), and, to be honest, Payette didn't offend me in the way Sheer meant. Unlike some members of modern society who seem permanently offended, I do not consider offending someone to be a serious wrongdoing. Moreover, I've grown a thick skin over the years, and expect the militant faction of the atheist community to ridicule anyone who doesn't subscribe to their materialistic view of the universe. However, after cooling down and reflecting on things, I realized that Payette and Trudeau's actions had generated a number of reactions in me other than the reflexive Alex-Jones-style fit. These reactions led me to believe that they, on behalf of the Western secular establishment they represent, had committed a much greater wrongdoing than just offending people.

The first reaction I had was jaw-dropping disbelief that an astronaut and a former drama teacher-turned-politician, felt qualified to comment so conclusively on one of the most intractable areas of scientific research, and in the case of Justin Trudeau, dictate what the *truth* on the matter is. Maybe I'm wrong. Maybe they have read extensively on the science behind the origin of life, but if they had, then I am pretty sure they would have known better than to be so conclusive on this subject. The reality, I suspect, is that they

---

3   Alex Jones is the host of the radio show Infowars. Mr Jones is (in)famous for his angry rants (fits) at the establishment and the endless list of global conspiracies that he seems to uncover. Whilst it is tempting to go down the "Alex Jones" route when confronted by establishment shenanigans, this style may not be the most convincing.

had just unthinkingly acquired a meme peddled relentlessly by the academic establishment and mainstream media. If Payette had stuck to just astrology or homeopathy, the other targets of derision in her speech, they might have been on more solid scientific ground. But she didn't, and they aren't.

My second feeling was one of smoldering anger. These kinds of ill-conceived pronouncements by representatives of the secular establishment[4] have a chilling effect on discussions on topics which are far from settled, and they have implications on how believers are treated. They also stifle genuine scientific inquiry. I'm sure there were a number of scientists in that room who, like myself, have belief in a higher power, or intelligence, which operates outside of the materialistic realm of science. Like me, they may have spent years thinking about this subject, and, looking at the evidence, have come to a different conclusion from atheists. However, speeches like this would make it even harder for them to express or research their views, further enforcing the atheist dogma increasingly imposed by the elite.

The last reaction was to generate a fierce determination that I would do all in my power to put the record straight, and help others to do so, too. I know that many people of faith, especially if they work in education, healthcare, or academia, are made to feel as though they are idiots if they believe in anything other than this secular establishment's orthodoxy of a belief in naturalistic causes to life, the universe and everything. I believe that faith is under attack in a way that hasn't been seen since the dark days of the middle of the twentieth century, during which the extreme left in socialist and communist societies ruthlessly repressed religious belief. We may have less overt forms of repression today, but they certainly exist.

Trudeau seems like a sincere sort of chap, but he is a leader who is recognized around the world, and by openly supporting the public ridiculing

---

4   I mention the establishment/elite a lot. I am not anti-establishment per se, however, if I see the "ruling class" messing with the truth, especially one so important, then I cannot remain silent.

of those who believe that life was not just the result of a random process, he has given license to the kind of bullying and prejudice that is becoming common in all walks of life. This applies beyond the borders of Canada. I have witnessed it in my homeland, the U.K., and even on my travels in the U.S. I have met people of faith from all over the Western world who feel increasingly isolated, even persecuted because they believe in a higher power. I have given talks on the subject of science and faith, and people who attended these talks have told me how much they needed to hear someone speak plainly, but with authority on this topic. I hope the journey that we are embarking on in this book will provide the same kind of assurance and encouragement to a wider audience, as well as stimulate further inquiry to those who are just curious.

If you are not an expert on chemistry and biochemistry, then it is easy to feel intimidated by those who represent the political or academic elite who claim to hold a superior knowledge. I'm not intimidated, because while I am not an international thought leader, I have sufficient confidence in my knowledge of the relevant science, and in my powers of reasoning, to stand my ground, not least against an astronaut and a politician. Moreover, I know that the conclusions I have drawn on this subject are shared by numerous other scientists, some of whom are well-known and highly respected.

So what is the wrongdoing that our establishment representatives committed?

Violation of the truth.

Why not just say that they were misleading us? Why something as strong as violation of the truth?

Violation is a more powerful word and sums up the visceral reaction I had to this incident. It wasn't just the misrepresentation of "any old truth," it was much more than that. To me, the sacred purity of a central truth about scientific knowledge was perverted by their clumsy and ignorant statements. The truth that lies at the heart of this discussion is of the utmost importance and has vast implications for our understanding of the nature of reality. For the establishment to mislead people on this subject, whether deliberately or out of ignorance, is like vomiting into the source of a spring that provides water for the whole of humanity. The understanding of our origins is the basis of all knowledge about ourselves, and for the powerful to corrupt this spring of knowledge with misleading remarks is therefore a grievous wrongdoing worthy of a trial.

To discover whether Trudeau and Payette, and the secular establishment they represent, are innocent or guilty, we need to determine what the truth on this subject really is. What is the evidence, and what does it say? If the evidence strongly supports the idea that life came into being by random processes, and that to believe anything else is just plain dumb, then they are innocent. But if the evidence allows any room for people to rationally believe that life was initiated by intelligence, then they are guilty.

Before I introduce the arguments for and against random processes, it is worth understanding the history of the highly contentious divide that drove this wedge between the modern establishment and people of faith. It was only a few centuries ago that the "establishment" was the church and to question faith was a crime. In many Muslim countries this still applies, so in this text I am referring to the Western establishment. So what happened to change the core beliefs that lie at the center of the national and global institutions that form this elite? What is the path that led us from a society whose leaders repressed irreligious thinking to one that is almost the complete opposite?

1.2 AN EPIC ROW

Most people would view the emergence of the theory of evolution as the point at which science first challenged God's existence, and yet even Charles Darwin in a number of editions of *On the Origin of Species* allowed for the notion that the process of evolution may have had some divine input at the very start.

"There is a grandeur in this view of life, with its several powers, having been originally breathed by the Creator into a few forms or into one..."

It is possible, but not certain, that he didn't actually believe this, and that it was included as a sop to keep the peace with his Christian wife, Emma Wedgwood, and his own grandfather, another Christian, who was also a highly influential pioneer of the Industrial Revolution, Josiah Wedgwood. (Yes, Darwin married his cousin). Either way, Darwin, a genuine genius, had the humility to publicly concede that the answer to life's origins could lie outside of purely naturalistic means. It's a shame that the establishment pair, Payette and Trudeau, neither particularly noted for their genius, lacked this level of humility.

Irrespective of Darwin's sop, the row over the origin of life has now been going on for over 150 years, and it actually started at the dinner table of some relatives of mine. Given recent events in Canada and elsewhere, which show an increasing boldness on the part of the establishment to dismiss faith in anything but science, I feel that it's time that this Wedgwood help people focus once again on this intractable problem, and with the information we now have, see if we are able to make better informed conclusions about life's origins. But it won't be easy, the sides are deeply entrenched, a position that is the result of centuries of feuding.

While it's true to say that the publication of Darwin's theory of evolution by natural selection accelerated an already widening divide between the church and scientific communities, and was possibly the watershed moment, it was not the beginning of hostilities. The trouble really started in Italy in the 1600s when a hero of the Renaissance, Galileo Galilei, through mathematics and astronomy, confirmed Copernicus's model of a universe in which the Earth revolved around the sun. Galileo ended up on the wrong side of the Roman Catholic Inquisition, the pernicious establishment enforcer of the day, and was eventually placed under house arrest, where he died.

In our modern era in which everyone understands from an early age that the Earth is not the center of the universe, it seems odd that such an idea could cause so much trouble. However, back then, when the church was all-powerful, and any challenge to its authority on religious matters was a challenge to its influence, opposing the teaching that the Earth wasn't the center of everything was a very big deal. Galileo, by proving that the church had got it wrong for centuries on such a fundamental issue of science, undermined their monopoly on knowledge, and in so doing, challenged their right to lead. It was also seen as potentially challenging the teachings of the Bible, and therefore undermining the whole precept that the Bible was in fact God's word. Taken to its extreme, this thinking could lead people to believe that God didn't exist and that the whole Bible was made up. People like Richard Dawkins are the natural end-result of this thought path.

So the Catholic Church of the time feared that Galileo's evidence could subvert the very foundations of the faith upon which its empire was built and therefore resisted his ideas violently. They adopted this "resist at all costs" approach to a number of subsequent scientific developments after Galileo's, even opposing steam engines in the 1800s. This intolerance of innovation and inflexibility on certain questions of science led to the Enlightenment movement, which started around the beginning of the eighteenth century.

Radical Enlightenment philosophers and scientists, who initially met secretly in societies such as the Freemasons out of fear of persecution, understandably took the position that there was no place for the church in political leadership. Their teachings and philosophies formed the foundation of the French Revolution at the end of the eighteenth century in which the

French royal family and their supporters, the church, were forcibly removed from power, and their heads physically removed from their bodies by a guillotine. This eventually led to the modern, secular French state, in which the two powers of church and state are separate. Although due to different grievances, America's war of independence produced a similar outcome. In both instances, traditional establishments based on monarchy and religion were ousted by people sick of the abuses they had endured in the name of King and God. These old establishments were replaced by new establishments with a very different understanding of how government and the practice of religion should coexist.

The fact that the U.K. at the time was a Parliamentary democracy, with a figurehead monarch, and that the church had been put in its place centuries earlier by Henry VIII, was probably the main reason why there was no violent revolution to overthrow the establishment on the other side of the channel. However, the British were very much at the center of the Industrial Revolution, and whilst being spared the violence of its continental neighbor, were equally engaged in the debates at the center of the enlightenment--a revolution in thinking.

It is important to understand that these huge political changes were not due to a change in people's beliefs about the existence of a higher power. In spite of the huge leaps in scientific understanding and the sweeping political changes, prior to Darwin's discovery the majority of people, including the great scientists of that age, retained a belief in God. They just decided that organized religion should no longer be one of the main authorities when it came to governing, and over time created political structures that reflected this decision. In fact, many of the great scientists of the time, such as Sir Isaac Newton, pursued their research in the context of trying to learn more about God's creation. Even Charles Darwin himself started out with a desire to enter the clergy. In the early days of his voyage on the Beagle, a British navy survey ship on which he was posted, he set out to look for evidence for God's creation, and it was only when his research led him to understand that species were not fixed, a finding that contradicted the Biblical account of creation, that his faith became wobbly.

His discoveries led him to develop his theory of evolution, which he outlined in *On the Origin of Species,* which he finally published on November 24, 1859.

Once again, the church opposed this discovery, but by now was powerless to stand in the way of its dissemination. Geological sciences had already been pushing the idea that the Earth was millions, if not billions of years old, and this new theory, which was somewhat supported by fossil evidence, went right to the heart of Biblical belief and challenged the idea that species, humans in particular, were uniquely created. This fueled the increasingly bitter row between the church and scientists. Given the history of the relationship between the two groups, starting with poor old Galileo, and the eventual removal of governing power out of the hands of the church, it is understandable why so many scientists gave the church their middle finger. There was already bad blood, and therefore no love lost between the two groups, but now scientists could see that it wouldn't be long before they would have the whip hand, and no longer needed to tolerate insufferable intellectual subversion of the kind attempted by some churchmen.

The twentieth century saw science solve some of humanity's greatest problems (as well as create a few new ones). Scientists transformed our world from one that was lived at walking or horse trotting pace, to being able to cross the planet in a day, or circle it in an hour. Science freed ordinary people of the West from bondage to the land or to daily household chores. Science helped us go from a situation where it would take months for communication to occur between residents of far-flung countries, to being able to speak face to face in real time on a mobile phone anywhere. Science, and a science in which I have proudly played my part, saw life expectancy double; old diseases consigned to history; and new ones like HIV tamed. In one century.

Had the church done anything like that? How long had the church been around?

No wonder so many people have put their faith in scientists over preachers! This faith naturally extended beyond just the science that generated technology, but to scientists' claims about our origins and the existence of God. In spite of this, people need to remember the fractious historical context that might have introduced irrational bias into the mind of the scientific community and created this unnecessary divide.

Hatred makes it all but impossible to see another person's views as being correct, and due to the unjust way that the church initially treated scientists who valued truth more than religious doctrine, and the subsequent mind-numbing stupidity it has shown when challenged with indisputable facts, it is no surprise that such an enmity grew between church and science. This hatred has now cooled into a snobbish disdain, and the hot lava of rage has hardened to a granite-like dismissal of anything coming from the church with regard to scientific understanding.

From what I have seen, I am convinced this attitude has finally penetrated the heart of the Western establishment. Atheism immediately became the central thinking of communist and socialist establishments after revolutions in the East, but in the West, it's hard to pinpoint the exact overarching attitude of the elite because it is a spectrum, albeit less broad than found in wider society. However, in general, while it may not necessarily always be overt atheism, there certainly appears to be an increasing readiness to dismiss the relevance or existence of a "God". As a result, the establishment attitude is now at best ambivalent about faith, and at worst actively, but not openly, hostile towards faith…that was until November 1, 2017. Payette and Trudeau are a part of this establishment and their actions were both open and hostile.

The general impression of most people, propagated by the establishment and its mouthpiece, the mainstream media, is that scientists have won all arguments and answered all questions, including the question that lies at the center of our quest for truth. It is now being made clear to us that religious faith will only be tolerated provided it doesn't interfere with their objectives.

It would appear there are those that feel the time is right to go even further. Payette and Trudeau may perhaps be the vanguard of this new approach. Instead of the tolerance these types claim to practice, it feels like they would prefer to silence religious views in the public square that contradict the dogma that only science can answer the big questions. This may come from a sincere belief that this is the best course for humanity, but given the lack of a truly objective consensus on the subject (something we will examine in detail), they may be sincerely, and dangerously, wrong.

I sometimes wonder if influential atheist cadres within this modern secular establishment have generated the meme that science and faith are incompatible with an unstated goal to eliminate all faith-based thinking from those who lead or those who teach so that future generations of minds are "cleansed" of unhelpful religious thoughts. It certainly feels like it sometimes, and while it would be slightly unbalanced to believe that the whole of the establishment is working towards this aim, it is fair to say that outside of a few states in the U.S., they appear quite happy to stand by and let this attitudinal cleansing occur across countless institutions in numerous different countries.

While keeping formal, organized religion out of political decision-making has been accepted as good standard practice by virtually everyone in the West, it is quite another thing to attempt to remove it from the minds of individuals. That is Orwellian in every way...*1984* and *Animal Farm*. The behaviors being displayed now have a whiff of the Inquisition about them. It is completely wrong to try to eliminate spiritual understanding from the discussion of philosophical questions arising from scientific discoveries that are ambiguous in their implications.

The attempted silencing of dissent against this materialistic dogma by the elite is helped by the fact that most religious leaders are still scientifically illiterate, and therefore unable to effectively defend society against this creeping censorship. Unfortunately, they tend to resort to statements of faith. That is bonkers. Most people are not now born into faith, and many are increasingly moving *en masse* towards agnosticism, or atheism. Many scientists have completely rejected faith, so they just laugh in the face, or more likely, behind the back of someone resorting to faith-based arguments.

As a result of these changes that have occurred over centuries, and carry with them deep-seated emotional baggage, scientific research, which started out as a search for an understanding of "God's Creation," is now conducted in academic institutions cleansed of overt religious thinking. They employ the unofficial dogma of methodological materialism: there is always a naturalistic explanation for material observations. Given its apparent history of endless victories in the various arguments against the church, this may seem logical to the majority of people, and from a purely practical perspective, it forces humans to work harder to find solutions to the problems that face us.

However, human rights and freedom of speech issues aside, is this an objective position? Is the establishment right to sneer condescendingly at people who believe that life was initiated by an intelligent creator? Is the dismissal of such beliefs the result of rigorous, objective scientific analysis, or is it perhaps the result of ingrained prejudices handed down from previous generations of scientists who fought a righteous battle against an unjust and willfully ignorant church?

Ultimately, the most important question is whether the dismissal of religious beliefs is objective. If it really is objectively stupid to believe that a supernatural intelligence initiated life, then maybe holding those believers up to public ridicule, like Canada's Governor General did, is the way ahead (better than locking them up, as they still do in China). But if there is no objective reason to call these people stupid, then it is most definitely not the right way ahead. Moreover, if there is objective evidence to support their beliefs, that changes everything.

We need to get to the bottom of this. We need an objective understanding of what the relevant scientific data says about this central issue of where we came from and, therefore, who we are. It really matters.

1.3 THE DIFFERENCE BETWEEN BELIEVED (SUBJECTIVE) TRUTHS AND SCIENTIFICALLY PROVEN (OBJECTIVE) FACTS

A friend of mine recently asked me over a beer how I, a scientist, can still believe in God when all the science disproves his existence. Like most people, I suspect he relies on the less-than-reliable mainstream media for most of his information, and this suspicion was confirmed when I asked him, "Which science is that?" His silence emphasized the fact that he didn't actually know of any science that disproves the existence of God; he had just acquired this assumption from others. He had not himself looked at the evidence for and against either position in a rigorous way, and had accepted the commonly-held position on this: the one peddled by the establishment media. I will come back to the media in a bit and show why my wariness may have some justification. In fact, I suspect you will be so horrified by the example I provide that you'll start to understand some of the factors that may have given rise to the potential wrongdoing at the center of our investigation, and you'll wonder whether the forces behind it are actually based on any logic at all.

Anyway, not only had my friend exposed his ignorance of actual facts on this subject, but he also exposed another fundamental flaw that lies at the heart of atheism and methodological materialism. It is actually impossible to disprove the existence of God using science. Science improves our understanding of the physical realm; this may inform us on the literal accuracy of the parts of the Bible that refer to the appearance of the universe, life and human beings, but in spite of atheistic assertions, this is irrelevant to the question of God's actual existence. If scientific evidence does, indeed, strongly contradict aspects of Genesis, this only challenges the scientific accuracy of verbal accounts passed down through numerous generations before being recorded in writing by the Jews nearly three millennia ago. The assertion that this disproves God's existence is one of many conflations[5] used in arguments on both sides.

So what is the conflation that atheists are using in this instance? What are the facts that atheists may be conflating?

Now let's just say, purely for argument's sake to understand how atheistic conflation works (and before any fellow believers start calling for the burning of my book, I am not saying I agree with this position), that science had proven definitively that species had evolved by Darwinian evolution. This is, of course, what is actually taught in schools and accepted by most in mainstream science up until recently[6]. This then--only for the purposes of this argument --is "Fact" 1.

Fact 2: The writers of Genesis believe in God.

Are "Fact" 1 and Fact 2 related? Is the "fact" that science confirms beyond all doubt that species evolved related to the fact that a group of men from ancient times believe in God? No. But there is a third "fact".

Fact 3: The account of the appearance of species in Genesis contradicts the scientific theory of evolution, which **for the purposes of this discussion only** (just in case you missed that point earlier) is a proven fact, therefore the writers of Genesis were wrong on this subject.

Now comes the atheist conflation: Fact 2 (Genesis writers believe in God)

---

5   Conflation in arguments is the treatment of two different, unrelated, concepts as one. For example: saying that if you add an apple to an orange you get a banana. Conflations are commonly used by people who lack convincing evidence to prove an argument. If you add two related facts together, then you are actually "building" evidence to support an argument. When you add two unrelated facts together, you may give the appearance of building a case by increasing the volume of data, but it is a sleight of hand as the data is unrelated so not additive.

6   At a 2017 meeting of the Royal Society, one the most respected scientific forums on the planet, a serious schism opened up between Neo-Darwinists and those researching epigenetics who are starting to question whether the appearance of new species or phenotypes can be accounted for by the Neo-Darwinian process of stepwise mutations. This will be touched on in a bit more detail, but not much as I avoid the subject, in subsequent sections.

+ "Fact" 3 (Genesis writers were wrong about how species appeared) = there is no God.

That is a conflation, and the "facts" in this example say nothing about the existence of God.

People also conflate the question of God's existence with the issue of suffering. Here are the facts:

Fact 1: Believers say God is good.

Fact 2: Believers say that God is all-powerful.

Fact 3: (People say that) suffering is bad.

These three facts, and they are facts, are often added together to provide evidence that God doesn't exist. But it is just another conflation leading to a conclusion based on opinion rather than the facts cited. The only conclusion that might be correct to draw requires that you believe in God, but that he doesn't measure up to our understanding of good. To conclude that these facts combined disprove God's existence is a logical fallacy[7]. This language may sound a bit pompous, but it is precise.

Our atheist piñata, Richard Dawkins, is the king of conflations. In *The God Delusion,* he combines his argument that the theory of evolution disproves the Genesis account of creation with his observation that religion is responsible for all kinds of atrocities, to conclude that believing in God is a dangerous delusion. Adding an argument that does not specifically disprove God's existence to an argument that man-made religion is bad, does not for one second show that people are deluded to believe in a God.

It is really important to grasp this central fact when understanding the

---

7   A logical fallacy is an error in reasoning that renders an argument invalid. Also called a fallacy, an informal logical fallacy, and an informal fallacy. In a broad sense, all logical fallacies are non sequiturs—arguments in which a conclusion doesn't follow logically from what preceded it. https://www.thoughtco.com/what-is-logical-fallacy-1691259

stance of atheists: there is no scientific evidence to disprove God's existence. None. As I've shown, the evidence rolled out relates to specific scripture-based understandings of God's role in the physical realm, not to whether or not he exists.

Atheists do not, and actually cannot, disprove the existence of God. They would argue that this logic is false because we could say that it is also impossible to disprove the existence of a talking cow who is able to sing Elvis songs in Dutch and Greek while riding a monocycle. Both God and the cow are ridiculous impossible entities, the existence of which cannot be disproved.

The converse of that argument is that seeming ridiculous is not in itself *proof* that something doesn't exist. If an idea seems ridiculous that may either be due to our own subjective bias or it actually may be that the idea flies in the face of logic and evidence.

When it comes to the cow, our knowledge of cow intelligence, and our observation that they are not prone to singing or riding bikes, are universally accepted and provides very strong evidence that such a cow could not exist, and therefore almost certainly does not exist. On this basis, it would be wise to avoid making public policy based on the premise that such a cow exists…the construction of bike lanes for the cow being a good example.

However, when it comes to the question of the existence of God, the situation is very different. There are mountains of evidence that God exists, but with one tiny little problem…very little of it is "scientifically" measurable.

Outside of science, human testimony is used all the time to determine the ins and outs of various issues. Whether it be a witness in the court of law, a report by an explorer or (sometimes) the report of a journalist, we rely on other human testimony to make decisions. We apportion weight to their testimony about something we did not actually witness according to their character, relevant expertise and corroboration by other dependable witnesses. While no one has ever claimed to witness seeing a cow singing "Jailhouse Rock" in Greek on a monocycle, there are literally billions of accounts of people experiencing

"God" in varying ways. Not all of them are mad, and yet science, in general, has chosen to dismiss this vast sum of human experience as a delusional meme passed down the generations like a fairy tale. I believe it has done this not because of evidence, but because of its history with the church and the ensuing bad blood that I discussed in the previous section. A position based on enmity alone is not objective.

But science, for better or worse, does now rule the roost. Because of all science's great services to mankind, the keys to the "mind" of Western society have been snatched away from the church and handed to science. Because science decided not to allow human testimony as evidence on the subject of the existence of God, we are where we are…unless there is measurable evidence that scientists and the establishment are ignoring or subverting. We will learn if that is the case later. Either way, after this section, I will not invoke human testimony as evidence either. However, before we do get past this, I want to ensure that an important fact is fully understood: Atheism is a subjective belief system. To leave no doubt about this, let's go back to the cow.

Not only is there is no evidence whatsoever to support the fact the cow does exist (i.e. no witnesses), there is a lot of evidence against its existence in terms of our knowledge of cow intellect etc. While this alone does not prove that it doesn't exist, it means you'd be as dumb as a cow to believe it did and deserve the resulting derision or hospitalization that would no doubt go hand-in-hand with voicing such a belief.

With the whole existence of God issue, let's for a second say that there is no scientific evidence supporting his existence, but given what I showed previously, you'd also hopefully agree that there is also no scientific evidence to show he doesn't exist. So there is a difference already between this issue and the cow scenario. The balance of *scientific* evidence is zero in this instance. In other words, if it were the case that science had no evidence either way, why on Earth have we handed over the keys on this issue to science? Maybe we didn't. Maybe the church left them on the dining room table when it walked out in a huff after refusing to engage in reasoned debate, and science sneaked in and grabbed them before anyone else could!

Anyway, we are starting to enter a philosophical rabbit hole. Suffice to say, there is no evidence disproving God's existence and atheism is therefore, by default, a subjective belief, or worldview. It is more than a non-belief in God; it is a proactive decision, supported by erroneous conflations, to not believe the claims of billions that God exists. It is ultimately a belief system as much as Christianity, Judaism or Islam. (Maybe it would be more accurate to describe it a counter-belief system?) Atheist scientists like Richard Dawkins are fervent evangelists for their belief system in the same way that Billy Graham was for Christianity, and their passionate delivery of their message can be equally persuasive.

The atheist line of thinking extends beyond the boundaries of philosophical discussion into the broader pursuit of scientific research. Methodological materialism, or the materialistic method, the informal position taken by many in modern science, works on the assumption there is always a naturalistic explanation for every material observation. Strictly speaking this is not objective. This position is based on a *belief* that there is no force outside of scientifically defined natural laws capable of affecting nature. There is no actual evidence to disprove the existence of such a force, or of its ability to interact with nature or the material world.

As I said, for the most part this approach serves us well, but when the subject of scientific inquiry is whether life could have resulted from random natural processes or whether it required some external intelligent force, this lack of objectivity becomes a hindrance to genuine pursuit of enlightening knowledge. If you start with the position that there is only a naturalistic explanation, you will not be able (or allowed) to assess evidence for a non-naturalistic explanation.

In spite of its lack of objectivity, the dogma that there is a natural explanation for everything, the intellectual basis of the wrongdoing, is rarely challenged outside of churches, mosques, synagogues or temples. Because the stance between science and faith has become so polarized over the years, the atheism of many scientists has become deeply ingrained and self-reinforcing. Theists are put off pursuing a career in science, and even if they do, learn very quickly that if they are too open about their faith, then it may affect their chances of progression. As a result, fewer theists enter science, and when they do, rarely speak about their faith, meaning that this voice is not heard in academic settings anymore. This creates a feedback loop of self-censorship and exclusion of the potential discussion of non-naturalistic explanations.

Conversely, without the counterbalance of people of faith within science, those of a stronger atheistic leaning are left to run the show with little resistance, causing further polarization. This was, of course, overtly true in communist countries, and from what I have seen, has become increasingly the case in modern, Western academic institutions which have more than their fair share of hard-left atheists bent on purging religious belief from the minds of the next generation. I know they exist because I have met some of them. Given that most of the leaders in government and business etc. pass through these academic institutions during their formative years, this subjective viewpoint has now spread into the wider establishment where it poses an even greater threat to religious freedom of expression.

If this view is, in fact, not only subjective but actually completely wrong, then it must not go unchallenged. If we only offer banal, PC wrist-slappings, or statements of faith, the wrongdoings will only spread wider and become greater--history tells us that over and over again. That is why I believe our quest is so important. We must find out the truth and, if it contradicts the establishment's position, then we must speak it, loudly and precisely...while we still can.

I'm just going to share one more conflation with you:

Fact 1: Justin Trudeau and Julie Payette felt it was legitimate to mock people of faith.

Fact 2: Trudeau and Payette are members of the global establishment.

Conflation of Fact 1 and Fact 2 made by a science writer: The global establishment is bent on silencing all dissent against the mantra that science has all the answers and God does not exist.

It is of course possible that I, along with many others, are misinterpreting or conflating numerous events over the years, some of which we have experienced personally, to mean more is going on than actually is. It is hard to prove these things for sure. It is also possible that we are right. Personally, I would rather speak out and look stupid because I am wrong, than stay silent, and find out I was right when it's too late.

What about you?

In conclusion, science, contrary to the bluster of certain vocal atheists, has not disproven God's existence, and therefore it is rational and objective to allow for the possibility that there may be evidence supporting "non-naturalistic" causes behind life's origin. This is the baseline context that must be understood for anyone to make a truly objective analysis of the mystery surrounding the origins of life. If you assume it's not there, then you are unlikely to find it. Therefore, on our journey together, unrestrained by the atheist dogma of methodological materialism, we will seek to find out whether there is, in fact, empirical evidence to support or refute the existence of a "supernatural," intelligent involvement in the generation of the first life-form, and thereby determine whether or not a wrongdoing has occurred, and if it has, resolve to prevent any more.

1.4. FAKE NEWS

Whether or not my suspicions that the establishment is moving to silence faith-based dissent against the materialistic dogma is true or exaggerated, one thing is for sure, their mouthpiece, the mainstream media, is actively working to enforce the belief that the problem of the origin of life has been all but solved.

A central argument for the defense against the accusation that a wrongdoing has occurred might be the assertion by the media and college-level academia that science has already made great strides in uncovering the mystery of how DNA came into being, and that they are on the cusp of solving the problem entirely. No doubt Payette and Trudeau and many others sincerely believe this. The truth is, they are just as much victims of misinformation as the millions of others. They probably watched a news clip on the BBC, CNN or CBC that made just such a claim, and since they believe everything else that these "trusted" outlets produce, they had no reason to question them on this occasion, especially since the story was aligned with their worldview. I have news for them, such reports are invariably fake news.

There are many laws of nature and science that allow us to predict various events; one of these is an obscure law that I have discovered and named "Wedgwood's absurdity principle." This law, based on years of observation, predicts that at intervals of about 3-6 months various media outlets will produce headlines like "Chemists offer more evidence of RNA as the origin of life." Or on NBC the morning on which I started this chapter, "How space dust could help explain the origin of life." Or my personal favorite, from the U.K.'s *Daily Express*, "Origins of life SHOCK: Four billion-year-old rockpools are secret to man, say scientists." This is the newspaper that, every November, predicts that the U.K. will have its snowiest winter since 1963 and, every spring, predicts a BBQ summer akin to 1976, both events being once-in-a-century events. Like a broken clock, the *Express* will eventually be right in its weather forecast, but the same does not apply to the headlines about the origin of life.

The effect that these headlines have on the scientifically illiterate population is to create a sense that the origin of life puzzle is all but solved. I have seen the same kinds of headlines on the BBC, CBC and other mass media outlets, and just like an article on CNN about Trump, or on Fox about Clinton, I don't need to read it to know what it will say. I always do, though, and I always end up peeved that I wasted a couple of minutes of my life. Every time, without fail, the mainstream media have helped the increasingly atheistic establishment maintain the impression that great progress has been made in this area. But are they right?

Unfortunately, these headlines are invariably "fake news." The fact that they are misleading is clear as soon as you read them, since all of them (apart from the space dust headline[8]) describe how some chemist may have shown how it is possible to create suitable starting materials, or biological components of living systems (amino acids and nucleosides, which you will become very familiar with before long). What they never, ever show is how these might have come together in a way that actually initiates life through random natural processes – namely, how these components come together to form proteins or DNA, molecules which I will also introduce and discuss in detail later.

Here is a specific example of the type of fake news I am talking about. As of September 2018[9], if I Google "origin of life," the top link is to an article entitled "Researchers may have solved origin-of-life conundrum" which appeared in *Science* in 2015.

"Wow, them scientists sure are clever!" most people think.

Maybe they go on to read the first few lines, then get bored and search for

---

8   This relates to a theory called panspermia: that life, or the key ingredients were delivered on a particle from space. I will cover this theory in the appendix.

9   At the time of the final edit before publication in January 2019, googling "origin of life" gave a result list with the top 3 links being to the wikipedia page on the origin of life. This Science article discussed here, was 4th on the list, but was still the top "news" citation listed in the search. Refer to Appendix 8 and https://www.sciencemag.org/news/2015/03/researchers-may-have-solved-origin-life-conundrum.

the latest article on the size of Kim Kardashian's butt.

Job done.

What I mean by job done, is that the desired perception has been perpetuated. Most people never really get past the headline on these articles, as they either do not understand the material or they get bored. The headline is all that is needed to keep the meme alive. So what direct evidence do I have that this is an egregious example of fake news?

This article was reporting on research conducted by a chemistry lab in Cambridge University in England, led by a prominent origins of life researcher called John Sutherland. In 2017, Sutherland wrote a review in the highly respected science journal, *Nature,* on the state of the field[10] (origins of life research), including a summary of the research he had published in 2015 and which formed the basis of the *Science* piece. This is from the introduction, I bolded the most relevant sentence (spoiler alert: this may give you an idea where our search for evidence is heading):

"Understanding how life on Earth might have originated is the major goal of origins of life chemistry. To proceed from simple feedstock molecules and energy sources to a living system requires extensive synthesis and coordinated assembly to occur over numerous steps, which are governed only by environmental factors and inherent chemical reactivity. Demonstrating such a process in the laboratory would show how life can start from the inanimate. If the starting materials were irrefutably primordial and the end result happened to bear an uncanny resemblance to extant biology — for what turned out to be purely chemical reasons, albeit elegantly subtle ones — then it could be a recapitulation of the way that natural life originated. **We are not yet close to achieving this end, but recent results suggest that we may have nearly finished the first phase: the beginning.**"

---

10   Studies on the origin Of Life – the end of the beginning; J.D. Sutherland; Nature Reviews Chemistry, Vol. 1:12 (2017)

Does the last sentence sound like "Researchers may have solved origin-of-life conundrum"?

To emphasize his point, Sutherland created a graph as a visual demonstration of how far researchers have progressed in developing a theory of how life came into existence--namely, going from an early Earth environment with no biological chemicals to Life as we know it, or LUCA (Last Universal Common Ancestor, more on this later). This is my simplified version of the graph which captures its essence. (The line is slightly above zero at the highlighted point, but scaling makes it hard to see):

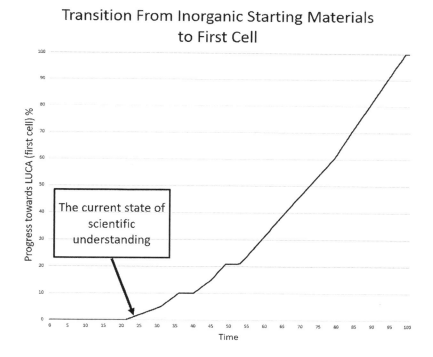

Transition From Inorganic Starting Materials to First Cell

Does this graph charting the progress of origins of life research suggest "Researchers may have solved origin-of-life conundrum"?

This is where my propensity for the use of "colorful" language is tested to the limit.

Just so we are absolutely clear, at the time this was written, if you googled "origin of life" the top news article appears to be verified Fake News. To put it politely, the headline does not seem to reflect the material it cites. This is especially egregious given that *Science* is a respected and trusted scientific news outlet.

Anyway, we need to move on before I have another Alex-Jones-style fit, and maybe this time never come down! But there is good reason to be angry; this is every bit as bad as the wrongdoing under investigation, if not worse, because it feeds the thinking behind the wrongdoing. This headline is blatantly polluting a sacred truth, the truth relating to our origins. This is the vomit in the spring.

To repeat the key point, since Sutherland and all other origin of life researchers do not answer the central questions of how DNA or proteins come into existence, they haven't even crossed the starting line in the quest to find a naturalistic explanation of the origins of life. It's a bit like saying that you have found out how to make a combustion engine, when all you have discovered is a source of iron ore. But in fact, it's usually even worse than that because not only do they avoid explaining the origin of DNA or proteins, but also they rarely show meaningful progress in how to make amino acids or nucleosides (the necessary starting materials). They either fail to simulate the prevailing early Earth conditions correctly, or use pure commercial reagents as starting materials, which are already complex and would not be readily available in an early Earth environment. We learn more about this topic, and whether even Sutherland's claims about making very modest progress are a true reflection of the state of the field, or whether they are in fact still at zero, in the section entitled "Getting to the Starting Line."

As an aside, I have some personal experience with the potential for hyperbole and hubris that this kind of research can generate. As I was preparing to get my Ph.D., in 1996, a section of our lab suddenly became very animated by a secret project. Three senior post-doctoral members of our group would regularly disappear into the office of Professor McGuigan, our supervisor, and then return to work until late on various experiments, completely unrelated to our usual work. After a while, the atmosphere relaxed a bit, and those in the inner circle could be seen sporting smug expressions and smiling to themselves and each other a lot. However, they had been sworn to secrecy about the reason for their smugness and were not tempted by bribes of beer in the local pub. Believe me, we tried.

After a week, Professor McGuigan held a meeting for all the staff and Ph.D. students. He announced that we were to have a major paper published in *Nature*, one of the most prestigious scientific journals, and it would describe a significant discovery in the area of prebiotic synthesis. This is the exact same area that all these fake news stories focus on--namely, the creation through natural random processes of biologically relevant molecules (amino acids and nucleosides) from non-biological precursors prior to the advent of life.

The materials that we worked with on a daily basis were, in fact, these amino acids and nucleosides that are used to create life's most important molecules, proteins and DNA, so if any lab was going to stumble across a discovery of this nature, it was ours. It turned out that one of the post-doctoral students, who was also working with these molecules, had generated a side-product from one of his reactions that had implications for prebiotic synthesis. The reason my colleagues had been scurrying around secretly was that they were repeating the experiments, and working on the paper for *Nature*, the same journal Sutherland's article appeared in. Once the paper had been accepted, the work stopped.

This was my first conscious exposure to "We've found the secret to the origins of life" hype and I was right at the center--well, maybe an inch away from it. This was also, coincidentally, very relevant for me personally. It was around this time that I first sensed the impossibility of life emerging from natural processes. My focus, then, was on this very issue--namely, the appearance of key biological components in sufficient quantities or purity for life to start. From my own experience of chemistry, I felt it was completely implausible, and that this presented an insurmountable barrier to the natural explanations. I had not yet come to appreciate that these were only initial hurdles, therefore this discovery had the potential to challenge the course of my thinking at the time.

We all went back to work and looked forward to the imminent day when the publication would bring fame to our lab. One thing to note is that most of the players in this, including our professor, were fervent atheists. I had already had numerous discussions around the subject with a number of them, mainly focused on the chemistry. This was obviously a period when I was a little less vocal on this subject than before, especially as the precise details of the discovery had been kept under wraps.

About two weeks before the edition of *Nature* was due to go to press, the atmosphere in the lab changed again, but not for the better. There was another uptick in activity. Once more, the players would disappear to have meetings and return to the lab and work till the early hours. Those of us who weren't in the loop, but just observers, watched with curiosity, and noted that the expression of smugness had changed to anxiety, and the looks they gave each other were nervous.

Then the end of the activity came, the mood had changed from anxiety to despondency, another meeting was held for everyone, and it wasn't good news (for them). The paper had been withdrawn. The only positive aspect was that it happened with just a couple of days to spare because, had it actually gone to press, it would have required a public retraction, which would have been excruciatingly embarrassing.

So why had it been withdrawn? My colleagues' findings had turned out to be false. The potential new origin of life theory had been based on the production of a vital biological precursor (more on what all this means later, but essentially amino acid or nucleoside, the components of large biologically functional molecules) as a side-product from a reaction between two reasonably common chemicals that might have been present in the early Earth environment. The team's repeated experiments had shown that their theory was correct, and therefore they felt that it was properly validated, until one day one of them performed an analysis of the supply of starting materials used in the experiments. It turned out that this was contaminated. When the material was sourced from a different supplier, the reaction did not work, and the thesis was destroyed. Oops.

I learned a lot from this little episode, not least to be very careful about making grandiose scientific claims without being very certain they are based on accurate information. Secondly, it reinforced my growing understanding of the huge chemistry-related barriers that must be overcome to get to the starting line of the quest to solve one of life's greatest scientific puzzles.

As can be seen from Sutherland's 2017 review, no new data has come to light in the subsequent two decades since this anecdote that would change my understanding. I will cover what meager evidence there is in the section titled "Getting to the Starting Line."

This episode also showed me the eagerness of some atheists to show that life could have come into being through random processes. It is the Holy Grail of atheistic science. By proving life began without intelligent input, you would really kill off belief in an intelligent creator.

However, the same goes the other way 'round. If it were established that life could only have come into being through intelligent initiation, then atheism is dead. Period.

Much is at stake, and everyone knows it.

This is why I believe the secular establishment, through its compliant mouthpiece, the mainstream media, "hails" each new completely irrelevant chemical theory as "Scientists have found the secret to the Origin of Life". They know most people are not sufficiently fluent in the science to understand that it is fake news, or are too lazy, or too wedded to an agnostic or atheist position to challenge it. So CBC, NBC, the BBC and so on get away with it, and each time they do the perception that scientists have cracked one of life's deepest mysteries is reinforced, thereby diminishing further the perceived need for a creative intelligence. This, then, becomes the prevailing view of the majority without ever providing real evidence to underpin it. It is pure, atheistic propaganda creating groupthink.

As stated and shown previously, the atheistic understanding of the universe is a subjective one, and if there is significant evidence to show that they may be wrong, then all of the behavior I have outlined by the establishment or the media associated with propagating and enforcing their worldview on the masses is a gross and heinous wrongdoing on an industrial scale. If that were true, then you and millions of others have been deliberately mislead, or even lied to.

How does that make you feel?

Not brilliant I suspect. There is a way to make yourself feel better though: get to the bottom of this. Experiencing the revelation in your mind of an important truth is a wonderful, liberating experience. It's like your mind and person expand a little bit, and you get a craving for more. But it gets better. You are equipped to make changes in the minds of others, and that is when you start to feel awesome.

If you read on (and, as I suggested read other, contradictory material too), you will learn more about the evidence for and against natural or intelligent sources for the origin of life and be fully equipped to act to put the wrong right…if a wrongdoing has indeed been committed. However, I suspect that having had a brief glimpse behind the curtain with this Fake News section, you may be getting the feeling already of where the evidence may be leading us.

We are now very close to beginning our thorough assessment of the evidence. We, unlike origin of life researchers, are *actually* near the end of the beginning! A few more introductory discussions are needed so that the staging is complete.

1.5 AM I SUFFICIENTLY QUALIFIED TO GUIDE YOU THROUGH THE EVIDENCE?

I'm not a world-renowned scientist on this subject like John Sutherland or Jack Szostak. The academic elite would likely view me as a "dollar-store" scientist. However, does that necessarily mean I'm not "fit for purpose"? Does my lacking fame mean I am not qualified?

Here are some facts about me to help you decide:

Firstly, I studied chemistry for my first degree in Southampton University, in the U.K. at a time when it was a national center of excellence for the subject. I then went to Wales to do a Ph.D. with an extremely gifted professor to look for cures for HIV, HCV and cancer. As a result, I have a doctorate in organic medicinal chemistry, which involved making new drugs. I wrote or co-wrote a number of publications associated with my work, the most important of which is cited below[11]. Interestingly, our lab made a very significant discovery (ProTide technology[12]) which has now been incorporated into a number of lifesaving medicines, including the genuinely revolutionary Hepatitis C cure, Sofosbuvir (Sovaldi™), which has already and, will in the future, save countless lives. Although I wasn't the individual who made the initial breakthrough, I feel privileged and proud to have been part of the team who developed its potential. Most of the time, research is boring and fruitless, but not in this case.

The specific focus of the research I undertook for my doctorate involved creating potential anti-viral drug candidates using nucleosides. Nucleosides are front and center in any discussion on the origin of life, as they are the key components of DNA. The other molecules that are central to any discussion on

---

11   Phosphoramidate Derivatives of 2′,3′-Didehydro-2′,3′-dideoxyadenosine (d4A) Have Markedly Improved Anti-HIV Potency and Selectivity; Christopher McGuigan; Orson M. Wedgwood; Erik De Clercq; Jan Balzarini; Bioorganic & Medicinal Chemistry Letters 6(19):2359-2362 · October 1996

12   https://en.wikipedia.org/wiki/Protide

the origin of life are amino acids, which are the components of proteins (I will explain the basics of DNA and proteins in a bit). As it happens, I used amino acids to improve the activity of the nucleosides. Much of my Ph.D. was spent synthesizing derivatives of the original lead compound, switching out different nucleosides and amino acids. I had to become an expert on the chemistry and biochemistry of these vital molecules. It was during this time that I really began to get a strong sense of how unlikely it was that such a system could come into being by purely natural processes.

Since finishing my Ph.D., I have worked in the pharmaceutical industry, sometimes in research and sometimes in commercializing drugs. I have worked in HIV, oncology and antibiotics. Many of the drugs I worked with use mechanisms that disrupt DNA replication, and so my familiarity with the biochemistry of DNA translation and replication has been maintained over the years.

Some of the roles I have undertaken have required me to present and discuss complex scientific data to different audiences, from junior nurses to top international medical experts. As well as learning how to express difficult science in understandable terms, I have also learned about credibility. As I point out in the opening section of the book, to be credible, not only must you have expert knowledge on the subject you are discussing, but any claims you make must be backed up by evidence and be balanced. That is my aim. I hope that when you reach the end, not only will you have expanded your knowledge, but you will also feel that you have been treated honestly and with respect.

Lastly, a number of years ago I faced a particularly challenging time in my journey of faith and I wanted to make sure that my beliefs were built on solid ground. Since I completed my Ph.D., I had always told people that I didn't believe life could have come into being by natural means, but I had never really proved it to myself. My belief about the origins of life was based more on intuition than on a thorough understanding of all the evidence and theories. So I set out to discover if any of the naturalistic theories for the origins of life held water, and whether there was measurable evidence for a non-natural, or intelligent source. I am lucky that I have access to primary, up-to-date literature from scientific journals, so I was able to be extremely thorough in my search. It's not unreasonable to say that I know as much about this subject, in particular the origin of DNA, as any other citable expert.

I hope my years of analyzing scientific publications, interpreting data and creating presentations, combined with my knowledge of the biochemistry and chemistry of the origins of life, will make you feel that I am indeed qualified to walk us faithfully through the evidence available.

However, even with the background and experience that I have, if I was the only person who held the viewpoint I have, I would at the very least have second thoughts about it. I might be confident, but I don't have the arrogance to think that everyone else on the planet is wrong. Thankfully, there are many scientists who share my understanding of this issue and have come to similar or identical conclusions. That not only gives me confidence, but it should give you some as well.

Suffice to say, even if you don't think I have the renown to speak with authority on this subject, I am only saying what others, who most certainly do, have said before. With that in mind, I will provide various materials from both atheist and theist scientists that are relevant to our task of uncovering the necessary evidence. There will be a few fresh insights that have not been articulated previously, which may help cement the final conclusions that we can draw but are not so radical as to change the state of play. The question is, what is the state of play? Has the establishment committed a wrongdoing, or are they right to dismiss those who believe life was initiated by causes other than random ones as slack-jawed yokels?

My goal is to ensure that, by the end of this book, not only will we have thoroughly investigated the evidence for and against the spontaneous appearance of life, but that you will also be equipped to articulate a position on the origin of life that is rock solid, evidence-based and balanced (i.e. not in the style of Alex Jones).

1.6 WHAT'S SO SPECIAL ABOUT THE ORIGIN OF LIFE AND SPECIFICALLY, DNA?

Humans are bugged by three big unanswered questions. How did the universe come into being? How did life come into being? And what is the source/nature of human consciousness?

In this book I focus on the origin of life, and specifically DNA for three reasons.

Firstly, I might have a Ph.D., but I recognize my limitations. I am good at understanding and communicating science, especially when it comes to areas I am familiar with like the chemistry and biochemistry of DNA. However, when it comes to some areas of scientific expertise outside of my own, in particular, theoretical physics, or cosmology, I am fairly ignorant. All I can say about the subject of the origin of the universe is that there appeared to be an origin, and that theoretical physicists have been able to account for everything except the first few nanoseconds of the Big Bang. I'm impressed with that, and I wish them luck in trying to prove that something comes from nothing, but I don't use this subject as evidence for God's existence. That is where I leave the subject. If Einstein can make mistakes[13] when it

---

13   Einstein's best-known mistake is a great example of what can go wrong when establishment dogma or personal beliefs influence scientific pursuit. Einstein's original calculations showed that the universe was expanding and, therefore, had a starting point, but this appeared to contradict the prevailing established academic consensus, or belief, that the universe had always existed and was static. Moreover, it would have supported theistic claims that God created a universe from nothing, and this went against Einstein's rejection of the belief in a personal God. To conform to the established scientific view and his personal belief system, he tweaked his equations, generating a number called the cosmological constant. This number was a patch applied to his mathematics that forced his equations to support a static universe.

Edwin Hubble, by discovering that distant galaxies were traveling away from us, showed that the universe was expanding after all, thereby confirming the validity of Einstein's original equations, and the fact that the universe had a beginning. The Big Bang. This story also shows the importance of realizing that established (and establishment) scientific positions can be completely wrong, especially taking into account their lack of objectivity as a community on the subject of God.

comes to this subject, then what hope do I have?

Secondly, while I am not an evolutionary biologist, I understand the principles of evolutionary biology but choose to avoid the controversy surrounding Intelligent Design (ID). ID certainly appears plausible (even obvious if you are a believer), but in my opinion, ID relies to some degree on the intuition that certain biological constructs appear to be designed. I know proponents of the theory argue there is more to it than that, and they are able to present good arguments that these constructs or systems have measurable elements of design, but I personally wouldn't feel on rock solid scientific ground supporting this position. I know that the "ID crowd" lump the origin of DNA in with all their examples of ID[14], but I believe that is a mistake. While the apparatus of translation (much more on that later) might invoke ID arguments, there is additional evidence of a different nature that is direct and quantifiable. In my view, ID is a rabbit hole that opens endless fronts that need to be fought.

Moreover, there is a strong[15], well-established, competing theory to ID that is not readily dismissed. Evolution by natural selection is a scientifically

---

14   Stephen Meyer, one of ID's leading proponents, has written an excellent book on the subject of the origin of DNA, *Signature in the Cell*, and the arguments he uses are similar to many I will use. Although I did my own research on this back in 2004, long before reading Meyer's book, and published my thoughts on the internet in 2008, the discussions on the following pages have no doubt been strengthened by reading Meyer's book and others.

15   One of the strengths of the theory of evolution is that when it was first proposed by Darwin, there was no evidence for genes (sections of DNA that code for specific proteins) or any means by which hereditary information might be passed from generation to generation, or how changes in characteristics that would confer advantage might be generated. Yet Darwin stated, without yet having the evidence, that this would most likely be the case, and that one day it would be revealed as the mechanism by which evolution could occur. That prediction, upon which his theory relied, was correct. In the next century we discovered DNA, how it is replicated and translated and that mutations do indeed occur. When someone sees new science and generates a theory from it, that is impressive, but when someone generates a theory and predicts the science needed to support it before the science has been discovered, then that is genius. This provides support to the validity of this theory.

rational and plausible theory. Yes, it is only a theory, but it is somewhat supported by evidence. ID is a new theory, but the evidence, in my opinion, is more subjective. That does not mean it is wrong, I just prefer to choose an area with strong objective evidence.

My third reason for choosing the Origin of Life and DNA, relates to what I just said. In my opinion it is the only area of science that has the potential to provide solid objective evidence of Intelligence. I will present this to you and you can decide firstly whether it is objective and secondly whether it helps us decide if the establishment is guilty of a serious wrongdoing.

1.7 WHAT DOES "EVIDENCE-BASED" MEAN?

This might seem obvious, but it is not always apparent that people are clear on what constitutes meaningful evidence, and how a position might be objectively formed on a subject using this evidence. There are criteria that we can use to define what evidence is relevant, and a method you can apply to weigh all the evidence in a balanced manner. We need to do this at this stage so that we take a consistent approach and therefore feel confident about any conclusions we draw after looking at all the evidence.

While many people will claim that their beliefs or opinions on any subject are based on evidence, some of us are clearly not objectively considering all the evidence, otherwise there would be no disagreements on anything. Why do I say that? Well, if all the available evidence for a subject has been considered objectively, then there should really only be one agreed, logical truth. In areas of emerging research or thought, where the evidence is not fully formed, there is argument over theories; scientists, journalists, and thinkers seek new evidence to shed further light on those theories. Ultimately though, to those who use logic, there is almost always one true answer to a question, even if you need to break that question down. This, of course, goes against the mantra of relativism, the doctrine that knowledge, truth, and morality exist in relation to culture, society, or historical context, and are not absolute.

Relativism is anathema to people of faith, and indeed to most people who use a process of logical reasoning to arrive at an understanding of something. To them, there is a right answer, even if we don't know it. Liberal arts philosophers who practice relativism will use religion as a case in point, and say that all are right or wrong, or none are right or wrong. Politicians who embrace this type of thinking try to impose this understanding as a solution to religious conflict. They would say that people's belief in God has evolved within, and is informed by, the culture they grew up in, and therefore, if there is a God, various people's understanding of him might be different, but all are valid. Religious people, for the most part, will disagree.

As a previous acquaintance once explained her relativistic view to me, religious beliefs are all "spaghetti in the head." She was a left-wing, atheist arts student, and the conversation didn't end well. Fundamentally, while I admire the desire to see everyone get along, and for disagreements to be resolved peacefully, and for all to be respectful of other beliefs, I believe that relativistic thinking is, for the most part, bad thinking.

If anyone was to ask what my greatest passion would be, I would answer in a heartbeat "the truth." This is probably why I have such a low tolerance for "spaghetti in the head." I want to get to the heart of an issue, whether it be data about a new cancer drug, or whether the establishment is misleading us on important issues, or whether life was created or just the result of a random process. I care passionately about the truth, especially when it's important, and I feel that relativism actively subverts truth.

So is truth always absolute?

Absolutely it can be. Often a question may not have an immediately obvious answer, but ultimately most questions about topics can be broken into a series of questions that have binary (yes/no) answers. These are truths. We may not always have enough evidence to state with complete certainty what the answer is, but that doesn't change the fact that there is an answer. In other words, we can state what we believe the truth to be but need more evidence to confirm what the truth is factually. The more evidence supporting our "belief" of a truth, the more likely that belief is the Truth.

An example is the question "Does God exist?" There is the binary answer – yes (theist) or no (atheist), one of which represents the correct answer as an absolute truth. Then there is the relativistic answer – God can be so many things to so many people so we can't define God. Spaghetti.

People's perception or understanding of who God is may differ, but if God does exist, people's views of who God is are irrelevant to who he (or she or it) actually is. This is why I feel atheists are more honest than those who adopt the intellectually lazy and jumbled relativistic approach. I can remember watching an interview with Richard Dawkins, and it was clear he is passionate about the truth. Of course, it may be that he is so blinded by his beliefs on this subject that if the truth poked him in the bum with a sharp stick, he still wouldn't recognize it. The same could, of course, apply to me! As mentioned in the opening section, psychologists have shown that it is very difficult for either of our brains to accept facts that challenge our worldview. That doesn't change the fact that one of us is right and one of us is wrong; there is a factual truth, even if neither of us is ever able to definitely prove it.

Ultimately, while it is currently impossible with the evidence we have to scientifically prove whether God exists, there is still a precise yes or no answer to the question of whether he exists. Due to the lack of clarity and depending on our worldview, we will give different weight, or bias to the evidence for and against.

What about the problem that forms the subject of this book, the origin of DNA? This has an absolute answer. There is only one true answer to the question "Did DNA come into being as the result of random natural processes?" and only one true answer to the question "Was DNA the result of intelligent initiation?" The two answers, in both cases, either yes or no, are mutually exclusive. If one is yes, then the other must be no.

Theistic evolutionists, or the BioLogos bunch[16], would say that it is not as simple as that. They would argue that God created all the laws of the universe, and life arose through natural processes as a result of these laws: so while God did not specifically create life, he created the laws that resulted in life, so ultimately he did create it. This is a fudge. However, to un-fudge the situation (always something that is needed once relativists get their hands on a problem), we must be more specific in the framing of each question: "Was DNA the result of random natural processes, as governed by the laws of the universe as we know them from observation of the universe, or not?" or "Was DNA the result of a specific act of intelligent initiation after the start of the universe, contrary to our understanding of the natural laws of the universe as we know them, or not?"

Put simply, did it happen "naturally"? Yes or no? Or did it happen "supernaturally"? Yes or no.

So the problem is not that there is no clear answer to a question, it is just asking the right question; then you can talk about what evidence is needed.

In a court of law, a judgement is made by weighing the evidence for either side of a case under review. Sometimes there is direct physical evidence relating directly to the case made for one side of an argument, making a verdict much easier. For example, in a murder case if fingerprints of the accused are found all over the scene of the crime and the murder weapon, this makes the case for the prosecution easier to prove.

Sometimes there is limited, unreliable, circumstantial evidence, making a decision impossible, and so the case is dismissed. For example, if a drug addict

---

16   BioLogos is a movement founded by the world-renowned scientist, Francis Collins, a physician and geneticist, who led the team who mapped the human genome, one of the greatest scientific endeavors of the last century. In his book, *Language of God,* which is the founding text for the movement, he discusses much of the evidence that we will be discussing here but reaches a conclusion that I am far from alone in finding unsatisfactory. I discuss this at the end of the book.

who is himself a convicted felon saw the accused in the area at the time, and there was no other evidence, his testimony might be viewed as unreliable and the case might be dismissed.

What is more common is when there is a mixture of circumstantial evidence for and against that, on balance, points towards one outcome. Circumstantial evidence could be that the accused bought a gun similar to the one used in the murder a few days before the murder, but the gun wasn't found. Or that a reliable witness saw the accused walking in the direction of the scene of the crime with a large bulge under their jacket…and so on.

A process is then applied to weighing this evidence.

I will use this same kind of process to help in the case of the origin of DNA in order to maintain objectivity. I really will try my hardest to be objective, and I will explain thoroughly why I discount or include evidence.

As you may have already gathered from the quote from Sutherland's review of the subject, there is not a whole heap of evidence for random causes out there. Some of the more detailed explanations and science may be put in the appendix section to avoid bogging down the discussion, but I believe that by the end of this book, main section, and appendix, all relevant evidence will have been discussed. It is worth remembering from the fake news section that new ideas are constantly being thrown out there, so by the time this actually gets published, there will be yet another headline "hailing" scientists making a new breakthrough to solve the origins of life problem. Maybe, once you have finished this book, you may raise a skeptical eyebrow as to whether this is real news.

I am going to create a table to keep score as we go through this process to maintain at least some objectivity with regard to the balance of all relevant evidence.

There are two sets of evidence to help us judge whether the establishment's position is a violation of the truth. One set is for or against random processes, and the second set is for or against intelligent initiation. This table is the starting point, a bit like John Locke's Tabula Rasa (blank slate)[17]. It would be good if you could emulate this in your mind, and wipe away all preformed ideas or opinions and just consider the evidence available:

	RANDOM PROCESSES		INTELLIGENT INITIATION	
EVIDENCE	FOR RANDOM PROCESSES	AGAINST RANDOM PROCESSES	FOR INTELLIGENT INITIATION	AGAINST INTELLIGENT INITIATION
AMOUNT (0-10)	0	0	0	0

Let's imagine we have completed the process, and the scores are as follows:

	RANDOM PROCESSES		INTELLIGENT INITIATION	
EVIDENCE	FOR RANDOM PROCESSES	AGAINST RANDOM PROCESSES	FOR INTELLIGENT INITIATION	AGAINST INTELLIGENT INITIATION
AMOUNT (0-10)	8	3	2	7

An objective assessment of the results shows that the balance of evidence relating to random processes supports the hypothesis that life was the result of random processes, and that the balance of evidence relating to intelligent initiation, does not support the hypothesis that intelligence initiated life. Random processes is the clear winner, and the Western establishment has good cause to promote the belief in anything else as being deluded. In this

---

17   John Locke, a well known English philosopher from the seventeenth century, theorized that the mind was a blank slate at birth and that all beliefs were later written on this mind by others. It would be helpful to try to create, or imagine that mental state with regard to this subject from this point. Imagine you know nothing, have no preconceived ideas, or beliefs relevant to this subject. I will return to this at the end.

instant the establishment will be found innocent. If the numbers had been the other way around, an objective assessment would conclude that the opposite is true. What about:

	RANDOM PROCESSES		INTELLIGENT INITIATION	
EVIDENCE	FOR RANDOM PROCESSES	AGAINST RANDOM PROCESSES	FOR INTELLIGENT INITIATION	AGAINST INTELLIGENT INITIATION
AMOUNT (0-10)	8	7	2	1

This is tricky. While there is more evidence relating to random processes, the balance is the same for both: +1. It's a tie, and you would have to concede that more evidence is needed to make a conclusive decision.

	RANDOM PROCESSES		INTELLIGENT INITIATION	
EVIDENCE	FOR RANDOM PROCESSES	AGAINST RANDOM PROCESSES	FOR INTELLIGENT INITIATION	AGAINST INTELLIGENT INITIATION
AMOUNT (0-10)	5	4	0	0

This example is not straight forward, but in general it is considered reasonable in this type of scenario, given that the balance of evidence just favors the theory that life could be the result of random processes, and *in the absence of any evidence favoring a competing theory*, to conclude that life was the result of random processes.

I can think of another area of disputed science that is going through a process of weighing evidence that is not dissimilar to this, with the same level of religious fervor displayed by the believers on either end of the spectrum of "belief." Climate change.

On one end, you have the true believers who will shout down anyone who suggests that the evidence does not point one hundred percent to the "fact" that manmade generation of CO_2 is entirely responsible for recent global warming.

On the other end, you have the sceptics who refuse to accept any associative evidence that correlates the rise of global temperature with the increase in industrial activity and CO_2 levels. The stakes again are very high. A leading group of climate science researchers made a statement on the sum of the data in 2017, which shows some of the logical processes that were used to weigh the evidence for and against human responsibility for global warming that draws parallels to what I describe above. This sets a precedent for the approach we are taking here…if it's good enough for an international panel of scientists, it's good enough for us.

So what constitutes evidence when trying to establish what was behind the appearance of life and DNA?

Is there any direct evidence? No. Obviously no one was around to witness life coming into being. All evidence in this case is circumstantial, and we will be considering the following types of evidence:

- **Geological and biochemical fossils.** This is physical evidence of historic processes found in the ground, and in living cells. I will give a weighting of 2 as it is as close to direct evidence as is possible in this case.
- **Laboratory simulations.** These are attempts to mimic "natural" circumstances. Depending on how well they correlate to our knowledge of the early Earth, and the ability to produce the required amounts of material, these get a weighting of 0.5-2.
- **Plausible theories supporting either hypothesis.** Without evidence, these theories get a rating of 1, with evidence, they get a 2.

To this last point, there are lots of theories that try to answer these fundamental questions. Some are worth considering, but some, in my opinion, are completely bonkers. For a theory to impact this discussion it must be plausible or viable. It's important that we establish some ground rules on what viable means. These are not formally established rules laid out by a scientific society or in a journal; they are just common sense. For a theory to be viable:

1. It must obey our understanding of the various natural laws that have been generated in the scientific disciplines over the past few centuries. These include the laws of thermodynamics, our knowledge of chemistry, and biological laws. It has to make *scientific sense*.

2. The statistical probability, if calculable, of it occurring must lie within the realms of known reality. To be more specific, if it is unlikely that it could have happened in the known universe, in the longest allowable time frame, then it is unlikely to have happened at all. There are some who will object to this and say we should account for a multiverse. There is less evidence for the existence of a multiverse than there is for a God...faith does not belong in this discussion. It has to make *statistical sense*.

3. It must be relatively complete. It would need to provide a logical, sequential path from the starting line to the current DNA/protein-based system of life. There may be some bits that need more data or are unresolved, and the odd section that needs more work, but there can be no huge glaring gaps, especially if they are conceptual. (If you can't conceive of it, then you are in real trouble.) Moreover, each step must obey rules one and two above. So if you had a path with 12 steps from A to Life as we know it, or LUCA, the following might just pass muster:

A > B > C> D – chemistry needs refining > E > F > G concept needs refining > H > I > J > > K > LUCA

In other words, it has to make *conceptual sense*.

To be viable, it should make scientific, statistical and conceptual sense.

The rest of this book will focus on all the available relevant evidence supporting, or refuting, the competing theories for the origin of life, and specifically the origin of DNA, so that you can produce an evidence-based "elevator statement[18]" summarizing our findings. After briefly covering the not-insignificant issue of getting to the starting line, I will focus on the three central problems that need to be solved to determine whether random processes could generate life.

- The numbers problem. They are really big, but are they too big?
- The chicken and egg problem. It's old, but is it rotten?
- The frozen accident problem. Is the DNA code really frozen, and if so, is it an accident?

In that last section, we will discuss whether any evidence presented for the case for intelligent initiation is real and scientifically measurable. While I've tried to keep things as simple as possible, it's a complex subject, and you may not "get" at least one of the sections the first time. (I've been told the frozen accident is easiest to understand.) Don't worry too much if you don't get them all; one or two will suffice.

Before we start, I want to briefly introduce you to a couple of friends, ones that you are intimately familiar with, but maybe never really got to know before now. This is the most basic of introductions, but you will get to know them very well by the end of the book. If you already have a good understanding of biochemistry, then skip or skim read these knowledge upgrades that I include from time to time.

---

18   Example: If you were a screenwriter, and got on an elevator and found yourself standing next to Spielberg, how would you pitch your idea in the time it takes to reach the 40th floor? 30 seconds to summarize the key points and make it convincing.

1.7.1 KNOWLEDGE UPGRADE: PROTEINS AND AMINO ACIDS – BRIEF INTRODUCTION

In the pages ahead, I am going to be using the words amino acid, protein, DNA and nucleoside a lot. As we work through the different problems facing random processes together, I will gradually increase the level of detail, but only as much as you need to understand the key concepts. The image of a protein below may look terrifying, but it is easy to understand from a conceptual point of view.

Most people associate the word protein with meat or eggs, and while these foods contain proteins, that understanding doesn't help here. Proteins are, in fact, very precise, tiny functional structures that have been assigned unique tasks that help build and maintain the organism they serve:

- A protein is a linear chain of 150 to 1000 amino acid (AA in picture below) molecules[19].
- Once constructed, this linear chain folds into a precise, three-dimensional structure.
- The type of amino acid (there are 20 to choose from) and the sequence in which they are joined together, determines how the finished chain (protein) folds.

---

19   A molecule is a unique grouping of different atoms joined together by bonds. Each molecule has unique shapes and properties.

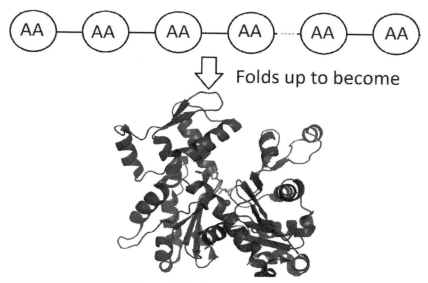

Folds up to become

SG Image of actin by Thomas Splettstoesser - Own work, CC BY-SA 3.0,
https://commons.wikimedia.org/w/index.php?curid=590649. Addition of AA chain made by Orson Wedgwood

For example, actin, a protein comprised of a linear chain of 374 amino acids[20], folds into the squiggly shape in the diagram above. It is a common protein found in muscle tissue and constitutes a high percentage of a blob of meat when you look at it but, given just how detailed and intricate the structure of a protein is, what we see when we say meat does not do proteins justice. It's a bit like looking at a Ferrari and saying metal.

If at this point you are getting concerned about understanding this, don't be. We will be visiting this again in much greater detail. However, I didn't understand everything about these things on first reading, so if you really want to understand everything, my advice would be to go back over stuff to really get your head around it. Go online too; others may explain it differently and help clarify things. Or you could just not bother… but remember, these molecules and the processes surrounding them are

---

20   In rabbits anyway…precise lengths may vary between species

central to our existence and understanding them will help you understand our origins.

For now though, all you need to remember is that **amino acids are the biological components of proteins.**

1.7.2 KNOWLEDGE UPGRADE: DNA AND NUCLEOSIDES – BRIEF INTRODUCTION

Most people associate DNA with the picture of the double helix that follows. In fact, DNA, just like proteins, consists of linear chains. The difference with the DNA **double** helix is that there are two chains spiraling around each other. The biological components of the chains of DNA are molecules called nucleosides[21]. These are linked together by a phosphate group (P in the diagram below). One strand of human DNA, which is the entire "code," or genome, for constructing and maintaining that human, is about 3 billion nucleosides long. There are only four different nucleosides available to construct these chains (much more on this later).

The sequence in which these nucleosides are joined forms a very elegant, and yet simple code. Segments of this code determine the sequence of the amino acid chains that make up the proteins from the previous page.

For now, all you need to remember is that **nucleosides are the biological components of DNA.**

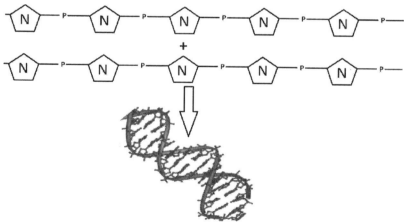

DNA double Helix image created by Jerome Walker and Dennis Myts; made available by Wikipedia

---

21   Purists would say nucleotides because of the phosphorous, but for simplicity, I will use nucleoside in this text.

KNOWLEDGE UPGRADE 1 SUMMARY

	Proteins	DNA
Core structure	Linear chain	Linear chain
Components of chain	Amino acids	Nucleosides
Number of natural components to choose from	20	4
Typical length	150-1000 amino acids	From a few million for a simple bacteria to 133 billion nucleosides for the marbled lungfish[22]
Appearance once chain complete	Tangled mess that is, in fact, a precise structure unique to each protein	Joins a "complimentary strand" and forms a double helix
Role	This "does" everything that is life	Source of information on organism. It "does" nothing. Like a book, it must be read to be useful
Connection	Proteins are coded for by DNA, but they also translate and assemble DNA	DNA chains contain segments within a genome that code for proteins. These segments are called genes.

---

22   Billion base pairs often referred to as Gb…this is analogous to computer information.

I have bolded what you need to know at this stage, but a few other things might stand out to you, and one in particular may have caught your attention… the difference in the choice of available components to choose from between the two systems. The reason for this lies in their function. This will be explored in detail, but just for a teaser now, four symbols can be used to say almost anything if they are combined using a specific code.

2. GETTING TO THE
STARTING LINE

2.1 WHERE IS THE STARTING LINE?

In this section, we are going to cover the problems facing Professor Sutherland, the researcher from Cambridge who produced the rather depressing graph summarizing the progress of origin of life research. He is joined by a multitude of other origins of life researchers who have focused on developing theoretical chemical routes by which the key components of DNA and proteins might come into existence.

Until you have a plentiful supply of these nucleosides and amino acids, you can't really think about actually building chains of nucleosides to make DNA[23]

---

23   For those familiar with the topic of origins of life, I will of course be discussing RNA later, and in great detail in the appendix.

or chains of amino acids to make proteins. No one is questioning the necessity of that prerequisite.

In the existing Earth environment, neither amino acids nor nucleosides form spontaneously outside of living biological systems. In other words, current natural chemistry, on land or in the sea, does not produce these vital components. They are either manufactured in cells by proteins or are acquired by gobbling the cells of other organisms. However, it is true to say that Earth's conditions were once very different, and that these components may have formed under these conditions.

There are a multitude of different proposals to explain how this might be achieved. Aleksandr Oparin and John Haldane were the founding fathers of this work, creating a framework back in the 1920s. It is interesting to note that both were fervent atheists and communists. Then came the famous Urey Miller experiments in the 1950s, in which this team generated an array of amino acids in their lab while using what was then believed to be a good simulation of the early Earth atmosphere. Those conditions have since been shown to be unlikely, but other attempts have been made over the years, most recently by Sutherland. I have summarized a few of these in the appendix.

They all have at least a whiff of validity about them, otherwise they would never have been published and obtained the fake news headline "Scientists have proof that man crawled out of pond 4 billion years ago" or whatever. Also, to be fair, these researchers are working in the dark, as there is no way of knowing for sure what conditions prevailed, or of validating their results, as the mineral deposits from those days do not contain evidence of any complex components. Because of this, no matter how valid their theory might be, it will always be preceded by the word "possible."

Given the plight of these chemists, toiling away in their smelly labs without reward, I think we should throw them a bone. Let's forget about all the complications like the incompatibility of the two types of chemical synthetic routes to get either amino acids or nucleosides. Let's forget about the complete lack of evidence in mineral deposits proving the likely presence of any of these components. Let's forget about the fact that the prevailing atmosphere and conditions were probably not conducive to any of these reactions happening. Let's instead walk along sunny paths and say it actually happened. Just for now, let's ignore the lack of evidence available, and say that, in spite of the all obstacles, somewhere sometime in the murky depths of the history of our wonderful planet, these reactions were able to happen and produce amino acids and nucleosides.

The question then becomes "Were these amino acids or nucleosides able to form in sufficient quantities and, equally importantly, sufficient purities to be of any use[24]?" This is, in fact, a more important question than whether a reaction could have happened. To understand this and get an idea of whether we can even get to the starting line, we need to consider two key factors:

1. Chemistry – what does our understanding of the science of chemistry say about the likelihood of sufficient amounts of pure components forming?
2. Chirality – what does our understanding of chirality say about the likelihood of pure components forming?

There is another event (other than the formation of large amounts of pure components) that is necessary to have occurred either before we get to the starting line, or immediately after: the formation of a cell wall. The reason why that is so important will be discussed at the end of this section. But first, a subject close to my heart and which I spent countless frustrating days pursuing:

---

24   The reasons why such large amounts of pure material would be needed will become apparent in the following section. Moreover, for reactions to even occur you need high concentrations of reagents, which means large quantities, and if they are impure then they will cross-react with the impurities, creating a useless mess. This section will make that clear.

2.2 CHEMISTRY

At this point, we need to distinguish clearly between three groups of chemicals so that you know what I am referring to when I use each term:

1. Generally available non-biological (or inorganic) starting materials;
2. The refined biological (organic) components of DNA and proteins: nucleosides and amino acids[25]; and
3. The end-product: biologically functional molecules, the DNA and proteins themselves.

Below is a cartoon of where the various groups of chemicals lie in a synthetic pathway:

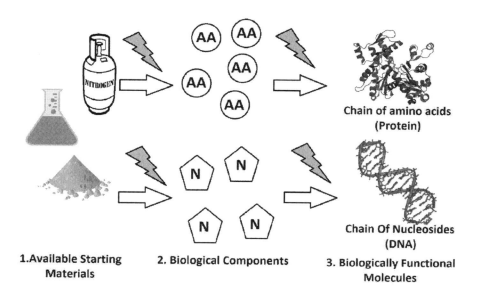

| 1. Available Starting Materials | 2. Biological Components | 3. Biologically Functional Molecules |

Chain of amino acids (Protein)

Chain Of Nucleosides (DNA)

---

25 This obviously applies specifically to a DNA-based world, which is the only system or biological "world" that we know has actually existed, but the same principles and requirements outlined in this section would equally apply to one of the various postulated precursor worlds – RNA, PNA etc.

A very rough analogy for these different groups of chemicals is to think of what is required to bake a cake. Let's imagine a perfect Victoria sponge cake, one that Nadiya Hussein of *Great British Bake-Off* fame would be proud of.

1. The raw ingredients are corn and sugar cane growing in the field, eggs in a chicken, milk in a cow, and strawberries; they need to be harvested, processed and refined before they can be used to make the cake. These are analogous to the generally available, non-biological starting materials.

2. The refined cake ingredients--flour, sugar, separated egg yolk and whites, jam, butter and cream--are ready to be used directly in the baking of the cake. These are equivalent to the biological components: nucleosides and amino acids.

3. The cake is the end-product, the biologically functional molecules; the fully assembled DNA or proteins themselves.

Generally available starting materials are ones that would have been present in the early Earth's environment. These could be in the atmosphere as nitrogen-containing compounds such as ammonia; or carbon-containing compounds such as carbon dioxide or methane; or materials on the Earth's surface such as water or phosphates. In addition, there would have been violent volcanic eruptions, huge electrical storms, meteorites and lots of upheaval...lots of energy, which is needed to drive reactions, but chaotic energy, which can also destroy products of reactions.

For random natural processes to generate life, it would be necessary for the generally available starting materials to form vast quantities of amino acids and nucleosides in high purity. The generally available chemicals would have needed to come together without the help of forced laboratory conditions, perfectly, countless times over to create these vital biological components. This is akin to someone jumping up and down in a wheat field and hoping to produce sufficient amounts of pure white flour for the cake. Just like with cooking, doing chemistry is not easy. In fact, chemistry is cooking for geeks, except if you get it wrong, the consequences can be more severe than a deflated sponge, and you could die from causing an explosion or inhaling hydrogen cyanide.

I have lots of experience with chemistry going wrong, even from an early age. As a kid, I was fascinated with chemistry and convinced my father to allow me to convert our garden shed into a lab. One day we were sitting with some guests on our patio on a summer's afternoon, drinking tea, and possibly munching on Victoria sponge. Suddenly, one of our guests asked if there was smoke coming out of the shed. The thought occurred to me at that point that I may have left a methanol burner on. Seconds later, flames were leaping from the roof. After a few minutes of adults running around frantically with hose pipes and buckets of water, the drama was over and, thankfully, the only damage was to the shed and my pride. I noticed afterwards that the tea had been abandoned for gin or scotch. Chemistry quickly runs amok if left to its own devices.

Suffice to say cooking and chemistry are not straightforward and require considerable skill and precision. As mentioned, my research required generating new potential drug candidates using organic, synthetic chemistry. These involved making nucleosides or amino acids, the very molecules under consideration, so this experience is directly relevant. Often, to create these I would have to embark on a sequence of reactions with maybe five or even ten steps. In each of these steps, I would need to generate perfect conditions for the reactions to take place. This is necessary standard practice when trying to generate relatively pure, complex molecules from starting materials:

1. Atmosphere: sometimes I would need to make sure there was no oxygen present, so I would flush the tightly sealed reaction flasks multiple times with pure nitrogen.

2. Solvent: the liquid in which the reaction would take place was critical to success, and in a multi-step sequence I would use a number of different solvents, often avoiding water, as it can interfere with reactions (the same reason for avoiding oxygen).

3. Sufficient amounts of pure reagents (reactive agents, or ingredients) to generate suitable concentrations, and avoid cross-reactions (reactions that you don't want).

4. Temperature: would have to be set to within a few degrees centigrade of the target.

5. The mixture had to be stirred nonstop for long time periods using automated stirring equipment. This was to ensure a constant, even distribution of the reagents and products.

6. pH: had to be right. Some reactions need an acidic environment, others alkaline.

7. Catalysts: some reactions required catalysts, chemicals that accelerate the reaction.

8. Stopping the reaction: the reaction would have to be stopped using specific reagents.

9. Purification: on completion, the mixture would need to be purified thoroughly to remove solvent, side products and unreacted reagents, often using sophisticated techniques.

If any of these conditions were not correct, the reactions would either not proceed at all, or produce tiny yields, or produce something that they shouldn't: a useless brown gunk, or tar.

Point 3, which related directly to the key question of this section, quantities and purity, doesn't just apply to the second step in this process--namely, the assembly of biologically functional molecules from biological components. It is also relevant to the first step. The availability to chemists of sufficient quantities of pure starting materials to use in their reaction sequences ensures that the experiments have the best possible chance of occurring as desired. In almost all of my experiments, I would have a plentiful supply of extremely pure starting materials (more than 99% pure). The amount of material is related to a field of chemistry called reaction dynamics, and the fact that many reactions are not a one-way street but exist in an equilibrium.[26] The purity is related to the potential for cross-reactions to occur. If a reagent is reactive, it is in a state of readiness to react with other molecules, not just the one you want it to. In my extensive experience with chemistry, impurities could (and almost always would) cause other reactions to take place and generate the useless brown gunk. Remember the origins of life near miss from our lab? Impurities matter.

Given how much precision and forcing of conditions is required to generate small amounts of pure chemical products in a lab, the theory, or (more accurately) belief, that the early Earth environment was capable of producing the large amounts of pure amino acids and/or nucleosides required to begin making proteins or DNA goes against all available knowledge that we have about chemistry.

Let's remind ourselves of the bone we threw our poor researchers and assume that an existing, or a future proposed pathway to amino acids or nucleosides, was possible at some point on the early Earth. If you take into account all that we have just learned about the fussiness of chemistry, the

---

26   Reaction dynamics is a topic that focuses on all the different factors associated with "driving" a reaction to one side of an equilibrium that exists between starting materials and products, or getting a reaction over an energy barrier. It is the job of a skillful chemist to manipulate conditions such that the reaction dynamics favor the desired side of the equilibrium. This includes a host of factors, which are related to the list of different conditions that I would need to consider when making a nucleoside or amino acid derivative. One of the key ones is concentration. One way of driving an equilibrium is to overwhelm the system with starting material. It also increases the chances of the right type of collision occurring.

most logical conclusion is that such a pathway would only have produced tiny amounts of impure substance. The same is true of the nutcase jumping up and down in the field. Something resembling flour might be found on the sole of his shoe but, as nice as the celebrity baker Nadiya Hussein appears to be, I'm not convinced she would take kindly to someone offering her the gunk on their shoe as a cake ingredient. The same goes for the mess that would have been created in the early Earth environment. Given that the kind of chemistry involved in generating amino acids, nucleosides, DNA or proteins, requires a much higher degree of precision to be successful than cooking, the proposed reactions, even if possible, would not have created sufficient amounts of pure biological components for the purpose of creating DNA or proteins.

Another explanation for the possible availability of biological components in the early Earth environment is that meteorites and/or space dust brought the amino acids or nucleosides with them. Aside from just shifting the same chemistry problems elsewhere in our universe, where an open environment would be equally unlikely to generate pure materials, there is the small issue of the fact that the meteorite would have smashed into the Earth at thousands of miles per hour. Moreover, as I have mentioned, there would have needed to have been extremely large amounts of pure amino acids to generate one useful protein, something we will discuss shortly. Such quantities would have needed to survive the violent impact as the meteor smashed into the Earth's surface, creating temperatures in the thousands of degrees centigrade, and a hot, miles-wide cloud of material. Believing this could have resulted in the delivery of large amounts of concentrated, pure amino acids is similar to dropping a bag of flour and some eggs from the top of the Burj Khalifa and expecting Nadiya to be able to use the resulting mess to bake her sponge.

Overall, the evidence from our understanding of chemistry is very strongly against there being sufficiently large amounts of pure amino acids and/or nucleosides present at the dawn of life on Earth to be able to form proteins and/or DNA (or any equivalent biologically functional molecule) from random processes. This is further supported by the fact that, while possible pathways have been mooted, no one has ever proposed how these pathways would result in sufficient amounts of pure components. Evidence for this lack of progress in even conceiving of solutions is highlighted by the previously cited 2017 review in *Nature* by Sutherland, an international thought leader in the field of Origin of Life research, working at the University of Cambridge, the globally renowned university where Watson and Crick discovered DNA. This is the same state of play that has existed since the 1950s.

2.3 WHAT IS CHIRALITY

Chirality (pronounced "kirality"), is a chemistry related topic, but is so fundamental to every aspect of biochemistry, and therefore the questions surrounding the emergence of life, that it deserves its own section. Before we are able to discuss why it is so important, it would help to understand what it is. Getting to grips with the concept of chirality will greatly enhance our appreciation of the difficulties facing the possibility that life arose due to random processes. If you know about this already or would just prefer to read the conclusions, then advance past this knowledge upgrade to the next heading (implications of chirality).

2.3.1 KNOWLEDGE UPGRADE: UNDERSTANDING CHIRALITY

Consider the following examples:

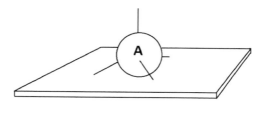

Figure 1

Imagine a ball with four rods poking out of it. These rods are equally spaced out on the ball. If you hold one rod loosely, the ball will dangle with the three other rods facing downward like a rotund tripod. You could put this down on a table (Figure 1: Ball A). You could then spin or turn the ball around by twisting the top rod with your finger; this would move the three rods on the table in a circle around the ball. If you turned the top rod a third, the shape would look exactly the same after the turn as it did before, as all four rods are identical.

If you rotated around the axis of one of the rods resting on the table (and allowed the other two rods to pass through the table surface) then you would still end up with what looks like exactly the same shape. No matter which way you rotate any rod by a third, you will end up with a shape which is superimposable with the image that you started with.

Now imagine if you put a circle on the end of the rod facing upwards (Figure 2: Ball B). Again, you could spin it a third around rod 1 in any direction and it would look identical after the spin as it did before (Figure 2: Ball C).

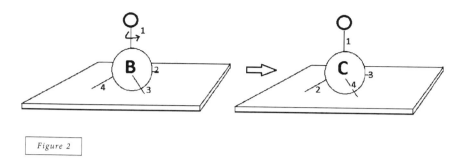

Figure 2

What if you put the circle on one of the other rods (e.g on rod. 2) resting on the table (Figure 3: Ball D)?

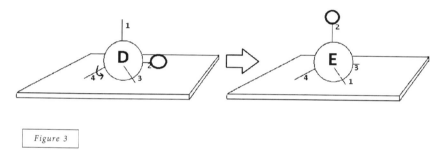

Figure 3

You could rotate anti-clockwise around the axis of rod 4 by a third and continue to move the circle to the original position at the top to make Ball E and it would look the same as i.e. be identical to, Ball B in Figure 2. The two images are *superimposable*. So if you add one circle to any of the four positions, the shape would, in all essence, be exactly the same; with a bit of twisting, the shapes are completely superimposable.

What if you have two shapes, a square and a circle, on the ends of different rods (Figure 4: Ball F)?

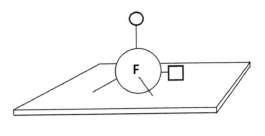

Figure 4

Does changing the position of the shapes on the ball (putting them on different rods) make the objects different?

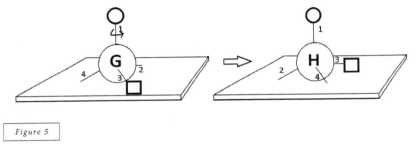

Figure 5

No. In the first example, Figure 5: Ball G, the position of the square is changed, but a quick anti-clockwise spin around the axis of rod 1 gives Ball H, which is identical to Ball F.

In the second example (Figure 6 Ball J), both the square and circle are in different positions, but twist anti-clockwise around rod 4 to give Ball K and you get back to the same original ball drawn in image F. They are identical.

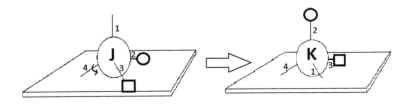

Figure 6

In fact, as long as two of the four shapes on the ends of the rods are the same (or as in this case, nothing), then it doesn't matter where you put them; the shapes are identical.

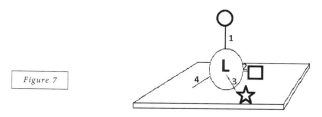

Figure 7

Once you have a third shape added, in this case a star (Figure 7 Ball L), does this change things? Does it still matter where you put the three different groups? (In fact, you now have four different things at the end of the rods, as nothing is different from something)

Below, in figure 8: Ball M, I have swapped the square and the star around from the positions in Figure 7 Ball L.

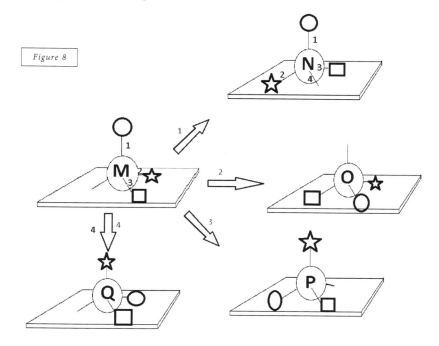

Figure 8

No matter what rotations you perform (in the example, we rotate the balls anti-clockwise one position around each of the axes), you will not be able to get the identically orientated ball and rod arrangement as seen in Ball L.

In fact, what can be said of these two objects (M and L) is that they are the mirror images of each other. If you rotate the table in Figure 9 Ball M to give Ball Q, you can see this mirror effect very clearly. So, while the balls have the same four groups on them (nothing, circle, square and a star), arranging them differently makes them different, and no matter what you do, while they are mirror (or stereo) images of each other, they are not superimposable.

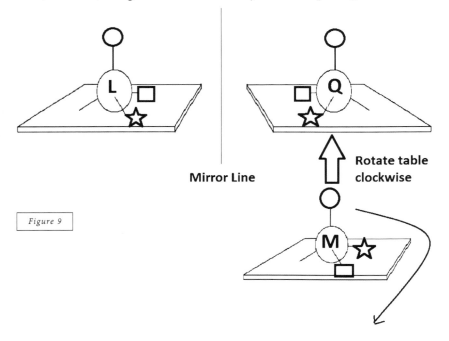

Figure 9

So what?

Carbon, the central atom in organic, or biological, chemistry is just like this ball. It can be joined to four possible different chemical groups and points equidistant around the central carbon atom (C). Methane is very much like the original ball, except you have hydrogen atoms (H) at the end of each rod (bond):

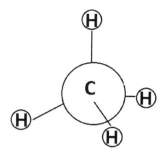

If you replace the hydrogen with a different atom or molecule through a reaction, then we are in the same position with the ball and rod example in Ball B where we only add one shape. You can put it anywhere and it will still be identical in every respect with the same properties (reactivity, shape etc.). If you add two chemical groups, then once again, you can create a superimposable shape by rotating. But what if you add a third group? As with the ball and rod example, they might at first glance appear the same, weigh the same, and have the exact same chemical composition, but they are in fact different molecules that often have different properties and react in different ways. A carbon atom like this is called a chiral center or "stereocenter."

Thalidomide is a horrible example of this. One of the mirror images worked to stop morning sickness, and the other caused horrendous birth defects, and yet they were identical in every respect except for the orientation around one carbon atom.

So what effect does chirality have on the question of whether random processes or intelligence were the most likely causative agents behind the appearance of life on Earth? How does chirality effect our ability to even get to the starting line?

Let's go back to the ball and stick idea. Imagine you have a ball with a circle and square on it already, just like ball L (a few pages back and below), and then you dip the spare or empty rods 3 and 4 in some glue. The ends are now sticky. Now imagine dropping this model into a box containing a star and a triangle.

You shake the box for a minute, until a loose shape has attached to each rod. There are two possible outcomes either image X or image Y:

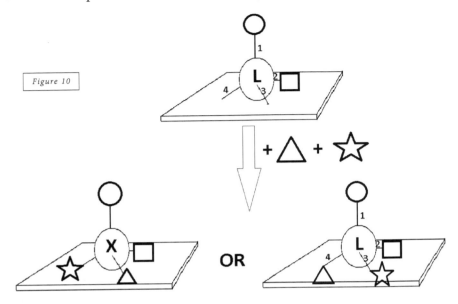

Figure 10

As explained, although they have the same number of squares, circles, triangles and stars, they are different, and the chances of forming either from shaking the box are 50%. There is a 50:50 chance that the star will attach to either position 4 first, or position 3, with the remaining position filled with the triangle. The same applies if the triangle attached first. In biology the two different orientations around the central atom are labelled either L (left handed) or D(right handed)[27]

In biologically functional molecules, it is vital that the components have the correct chirality (D or L orientation). Amino acids, the biological components of proteins, must be exclusively L, and nucleosides, the biological components of DNA must be exclusively D. Exclusively. The inclusion of a single biological component that is of the wrong chirality will destroy the effectiveness of either type of biologically functional molecule.

---

27   This is determined by the direction in which light is refracted when it hits them.

So how does this fit into the question of reaching the starting line?

L-alanine

Let's start with one of the simplest amino acids, L-alanine. A couple of things are worth noting:

- Where there is a meeting point of lines (which represent chemical bonds) and there are no letters (letters represent atoms or groups of atoms C=Carbon, H=Hydrogen, N=Nitrogen and O=Oxygen), that meeting point is always assumed to be a C, or a carbon atom (we lose the balls). There are two in this amino acid. The first is at the point represented by the symbol * and the second is at the point represented by #.

- Sometimes you see two (or even three) lines/bonds coming out of a carbon atom to just one other atom...these are called double or triple bonds. At point #, the carbon atom has a double bond to the O, or oxygen atom. When you have a bond like this, there are only three groups attached, the group lies in one plane (flat), and the opportunity for stereochemistry is lost.

- The wedged lines are meant to help you visualize what the molecule looks like in three dimensions. Imagine the solid wedges, like the one with the CH3 (methyl) group, coming out of the page (or screen) towards you, and the dashed wedges (which look like ladders), such as the one with Hydrogen (H), going out of the page away from you. The normal lines are in the plane of the paper, or screen.

The carbon center in this molecule at point * has four different groups attached to it. This is the only stereocenter of this molecule. In Figure 11 the balls are back to help with visualization:

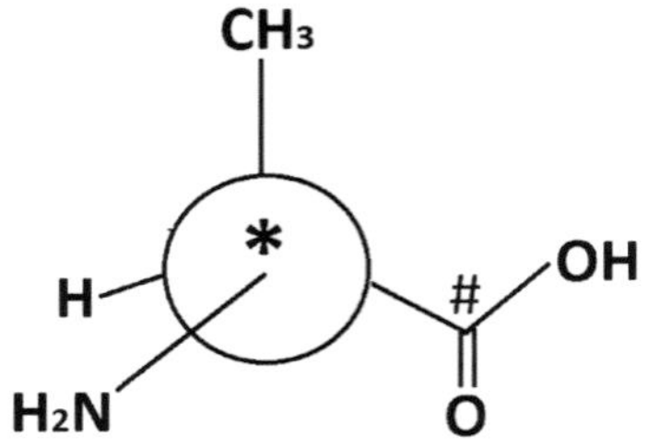

Figure 11

The molecule in Figure 11 is still L-alanine, I've just moved the methyl – CH_3 group a fraction anti-clockwise so that it is at the top like in previous ball and stick diagrams. But the orientation is the same as it is in the shorthand version. As you can see, this has the same "stereochemical" thing going on as the ball with four different shapes on it. The ball in this case is the central carbon atom of the amino acid (marked with *) and is the key "stereocenter" of this amino acid.

If you swap two of the groups on the carbon (e.g. H_2N with H), you create the mirror image, which is non-superimposable, and a completely different molecule. This is d-alanine:

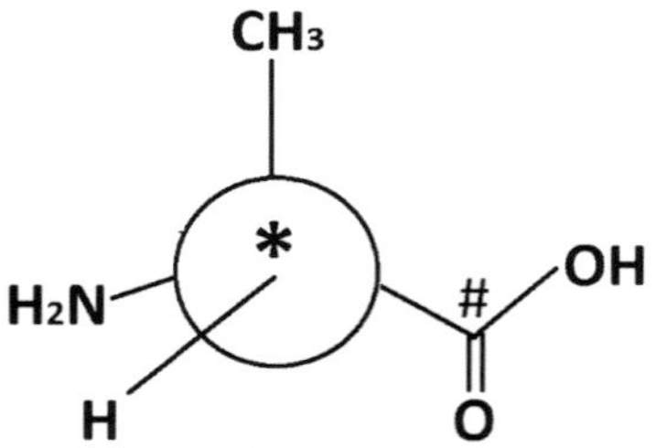

d alanine

2.4 WHY IS CHIRALITY IMPORTANT

To make biologically functional proteins (reminder, proteins are chains of amino acids), all the amino acids must be L-amino acids. As I mentioned previously, there are twenty naturally occurring amino acids used in assembling proteins, (I have put a table showing all twenty in the appendix[28]) but they all have the same amino acid "chemical motif," which is a central carbon atom (labelled with * in the diagram below) attached to 3 groups or atoms: a hydrogen atom, an amino (NH_2 or H_2N = same thing) group and a carboxylic acid group (CO_2H). The name amino acid comes from those amino and carboxylic acid groups. The group that changes between different amino acids is the group represented by the R in the picture below. In alanine, the R is a CH_3, or methyl group. In other amino acids, this group is different. All of these are shown in Section 2 of the appendix.

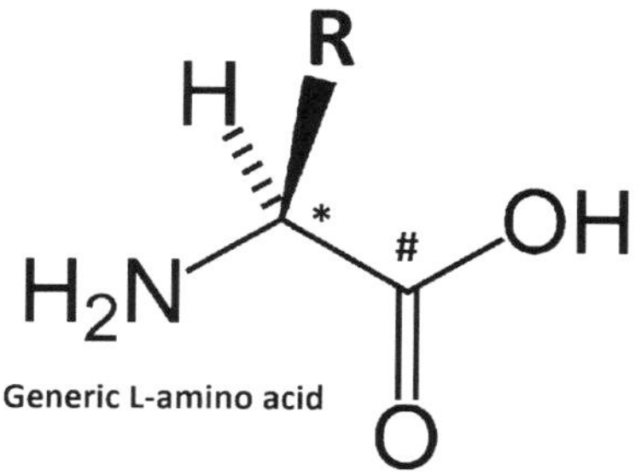

Generic L-amino acid

The one thing that must always stay the same to create functional proteins is the orientation of these groups around the central carbon atom. All amino acids MUST BE L, as shown above, not D (where the H and R would be swapped around). If random natural processes were responsible for the generation of proteins, the biological end-product, then there are two ways this could be achieved:

28   Appendix Section 2: amino acids

1. Either there would be a plentiful supply of pure L-amino acids, the biological components, with no D present, or

2. there would be a plentiful supply of a mixture of L and D amino acids, and as the protein chain grows, it only selects the L amino acids.

We are going to address the second scenario in the big numbers section, so here we will only discuss the likelihood of random natural processes producing large amounts of just L amino acids. Just as with the ball that has glue on the two available rods, there would be two equally likely outcomes in nature if random processes generated amino acids from an imaginary primordial soup, using the basic inorganic chemicals found in the early Earth environment. Such a process would generate a 50:50 mix of L and D amino acids. Each time a reaction occurs where a chiral center is produced, there is an equal chance of getting either L or D, and therefore random processes are exceptionally unlikely to be a source of pure L amino acids.

Therefore, even if the considerable obstacles outlined in the previous chemistry section had been overcome, random natural processes would be incapable of generating a pool of only left-handed amino acids, let alone generate entire chains of only left-handed molecules[29]. As I said, I will cover the assembly of proteins in the next section, but for now it is reasonable to conclude that the availability of "chirally" pure amino acids is extremely unlikely.

It is difficult to estimate exactly how unlikely, but for the sake of this section, and to get a taste of what is to come, let's say you needed to only make 150 amino acids to create the smallest viable protein 150 amino acids long. This is actually a massive under-representation of the number of amino acids that would need to exist as we will discover in the numbers section, but we'll run with it for now.

Should one of the potentially viable processes theorized by chemists actually

---

29   I discuss the Cairns-Smith "theory" in the appendix Section 1

work, these processes would randomly pump out L and D amino acids with a 1 in 2 chance of each one being L, in which case you would need to be "lucky" 150 times in a row.

This is the same luck that would be needed to throw a coin 150 times and get heads each time. The chances, or probability of this, are 1 in 2^{150} (2 multiplied by itself 150 times[30]) which would usually be expressed[31] as approximately 1 in 10^{45}. That number is vast--basically a 1 with 45 zeros after it. I will discuss big numbers like this, oddly enough, in the numbers section. All I will say here is that the chance of just 150 amino acids forming all L from a random process is exceptionally small, even in the 4.5 billion years that "Old Earthers" believe our planet has existed. If you were a super-flipper and flipped a coin 150 times every second for 4.5 billion years, you would still only have a 1 in 10^{28} chance of getting heads 150 times in a row. If you don't believe me, then give it a go.

The problem of chirality in randomly generated molecules is several orders of magnitude greater when it comes to nucleosides. These, remember, are the biological components of DNA. Below is a representation of the generic nucleoside used in DNA.

---

30   When you see a number in superscript like this it is called the exponent. It is the number of times that the number in normal size font is multiplied by itself. You will see lots of these, usually in scientific notation, which is the next footnote.

31   Scientific notation is the method of expressing very large number in powers of 10. A number is multiplied by 10^x where x is the number of times you multiply the number by ten, and therefore, the number of 0s after the number. 2^{150} is converted on a calculator to 1.43×10^{45} which, given the already silly scale of the number, is approximated to 1×10^{45} or 10^{45}

Nucleosides in DNA have 3 chiral centers at points 1, 3 and 4[32]. Each time a nucleoside is made and a new carbon center formed, you have two possible outcomes for each center. Since there are three centers you have 2^3 or 2 X 2 X 2 = 8 different potential outcomes for each nucleoside made. The idea that undirected, random processes, involving simple chemical constituents found outside of biological systems, could create a pure pool of correctly configured nucleosides is somewhat far-fetched. Even if you had a trillion years, it's about as likely as Hillary Clinton marrying Donald Trump. The next chapter is devoted to helping you gain a more technical understanding of what is possible and what is impossible so that you are not left too long with the thought of Donald and Hillary making out.

Remember, at this stage we are only talking about getting to the starting line. These are not the central problems of origins research, ones which baffle the entire scientific community. Maybe you are beginning to understand why scientists such as Sutherland adopt a gloomy tone when assessing the state of the field. He's not the only one, and I will cite others, equally well-qualified, who say much the same.

---

32   The modeling, labelling, lack of visible hydrogens, and reasons that points 1, 3 and 4 are all chiral centers are in the section of the appendix that looks at DNA and RNA structures. Also, in that section I discuss the issues with ribose chemistry.

2.5 THE CELL 'WALL'[33]

Another question that needs to be answered before getting to the starting line, or very soon after crossing it, is how did the cell wall come into being?

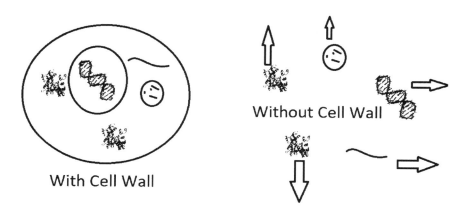

With Cell Wall

Without Cell Wall

All cells have a wall that separates its contents from the external environment. Each cell contains a collection of proteins, genetic material and various other bits and bobs required for fulfilling their functional purpose and replicating. All cells *need* cell walls. Each cell is a unique protected micro-environment with all the essential kit contained inside. Nutrients are allowed to pass through special channels in the cell wall; likewise, waste products or signaling products, like cytokines[34], are allowed to pass out. However, without the cell wall, the contents would spread out, dilute and become unstable in the surrounding liquid. In such an instance, there would be no means for the cellular machinery to work together and perform its assigned tasks. But how did a cell wall come into being?

---

33   For simplicity's sake, and to wind-up biologists, I use the term cell "wall" very loosely to include any structure enclosing a cell, including membranes.

34   Cytokines are very small proteins, usually less than 50 amino acids long. Strictly speaking, they are just polypeptides, as they lack active functionality like larger proteins. Their purpose is to act as signals. They bind to receptors on other cells, which then generates a cascade of actions. The receptors are the active proteins, the cytokines just turn up and present themselves. Cells will release cytokines to signal to other cells that action needs to occur.

The existence of the cell wall creates a chicken and egg paradox. Without a cell wall, cellular machinery can't work in concert, but the cell wall is created by the cellular machinery. (I will explain how the cellular machinery works shortly, but for now accept that it is the DNA and proteins that work together to do stuff that is vital to the survival and replication of the cell.) The instructions for building the cell wall are encoded for in the DNA of the cell, so how did the code for the first cell wall appear when a DNA/protein coding system (or something similar) could not work effectively in an open environment?

It is an important conundrum that must be solved before we can put together a viable theory for the origins of life. All life on our planet now consists of cells with cell walls. LUCA (last universal common ancestor) had a cell wall, so there is no avoiding this paradox.

If you wish to skip the next two knowledge upgrades, 2.5.1 and 2.5.2, move ahead a few pages where we pick up the cell wall conundrum with the imaginative title: Back to the Cell Wall Conundrum.

2.5.1 KNOWLEDGE UPGRADE: LIFE

For something to be "living," it must have at the very least the following three properties:

1. The ability to self-replicate/reproduce.
2. The information for its construction and maintenance are reproducible and passed on to the next generation (with the possibility for improvement).
3. To achieve survival and reproduction, it needs to consume and process nutrients (chemicals) from the surrounding environment to grow, reproduce and harness energy.

This could apply to a very simple chemical system, which many atheist scientists propose was the first living system[35]. The precise nature of this system has never been identified, or viably proposed.

2.5.2 KNOWLEDGE UPGRADE: FUNDAMENTALS OF EVOLUTIONARY THEORY

Overarching theory: Species evolve by changes in offspring that give them a survival advantage over the parent. This progeny, and its progeny, outcompetes the parent and other progeny of the parent that do not contain the advantageous change. i.e. survival of the fittest.

A silly, but effective, example of how it works:

---

35   Jack Szostak from Harvard is another leading origins researcher. His work on trying to generate the first self-replicating system is discussed in the appendix.

1. Blob A, the predominant blob in the ocean, is propelled around by the movement of water, and absorbs nutrients surrounding it.

2. Blob A produces millions of baby blobs. Some are just like itself; they consume nutrients and go on to live a happy life floating around the oceans. However, every now and then, the Blob replication machinery inside Blob A makes a mistake when it copies the code for Blob A. Most of these mistakes are fatal and the resulting baby Blobs die. However, one day, one of these "mutant" Blobs not only survives but has been given a code for fins. This is Blob B.

3. Blob B is now adapted to chase after nutrients much better than Blob A and all the Blob A type baby Blobs.

4. One day, there is a shortage of nutrients. Blob B and its offspring outcompete Blob A and its offspring for resources. Blob A starts dying off and Blob B predominates, or is **selected**.

Sounds great, doesn't it. In reality though, outside of my fantasy Blob world, fins don't just suddenly appear.

Neo-Darwinism relies on our knowledge of how information is stored and passed on through the DNA replication and translation processes (covered in the *real* chicken and egg section). It says that the change from Blob A to Blob B occurred through a stepwise series of beneficial mutations in the DNA of successive Blob progeny. Blob A.1 had a tiny growth that gave it a bit of movement advantage, Blob A.2 had a larger growth etc. until you get the fins of Blob B[36].

There are two fundamental laws that such a sequence of events must obey. Please imprint these in your minds, as I will refer to them a number of times:

---

36 Epigeneticists (and prior to them for the past two decades, the ID crowd), say that, from our knowledge of DNA, the appearance of a new phenotype like a fin by this type of process is not possible. I won't discuss these arguments, but some of their concerns are similar to ones I will use from time to time.

Fundamental Evolutionary Law 1: Natural evolutionary processes have no foresight.

- Natural processes, biological or chemical, are incapable of producing structures *purposefully*. They have no foresight or intention. They are not thoughtful, intelligent, goal-oriented beings.

Fundamental Evolutionary Law 2: For a mutant progeny to predominate, it must have been given sufficient survival advantage to outcompete the parent generation and its non-mutant progeny.

- In the case of our A series Blobs (Blob A.1 – A.2 etc.) the gradual changes in mutant growth must confer a survival advantage for them to become the dominant species.

- If something is successful at living, it has already adapted to survive and, in general, would need to make faithful copies for its continued survival (if it didn't, it wouldn't be well-adapted to passing on survival benefits). Mutations are mistakes, and most mutations reduce survival advantage. Because of this, large numbers of mutations passed from a parent to a progeny in a single replication cycle are fatal. Likewise, successive mutations in the progeny of the progeny would mostly be fatal or confer no advantage. If they don't confer sufficient advantage, they don't survive.

I give a real-life example from my years of working in the battle against the HIV virus and its ability to evolve that exemplifies these points in the 'Frozen Accident' section.

2.5 (CONT'D) BACK TO THE CELL WALL CONUNDRUM

Remember, in a self-replicating system without a cell wall, the rapid dissolution of the elements of this system into the surrounding environment would mean they would be unable to work together as a system, but without a working system, you can't make a cell wall. It's a chicken and egg paradox.

Various "compartmentalization theories" have been proposed for this, from oily bubbles forming fortuitously in proximity to the first self replicator and engulfing it, to the tiny pores in rocks being able to act as surrogate cells inside which cell like processes could begin to develop. Brian Cox, a celebrity scientist, is a proponent of the latter idea which he pushes in a clip on the Origins of Life that the BBC recommends to educators[37]. These theories do not answer the central question of how, or equally why, the code for a cell wall would ever be generated.

From what we just learned about evolution:

1. The How: A self-replicating chemical system of molecules would not have the foresight to look at the bubble and think to itself, "Wow that's a good idea, I think we'll code us a cozy wall so that we can get on with our tasks in privacy."

2. The Why: If a system had been compartmentalized in a surrogate cell wall, where would the survival advantage be in building a new one? In these scenarios, the need to produce a cell wall is removed if there is a "surrogate wall," so the evolutionary incentive is destroyed. Under a completely impossible statistical scenario in which one of the self-replicating molecules randomly produced the entire code for the cell wall in a few replication cycles, then it would not have a survival

---

37   https://www.bbc.com/teach/class-clips-video/the-origins-of-life-on-earth/zh8fcqt

advantage as it is already in a cell like environment. Moreover, how does it then get out of the bubble or the rock without breaking up?

Outside of the clearly unsatisfactory compartmentalization theory, there is a telling silence on the subject of the appearance of the cell wall by random processes.

The coding for a simple cell wall is many thousands of nucleosides long. (I will come on to how coding works later.) There are no credible scientists on this planet who would suggest that a new piece of working code of this kind of length could appear spontaneously through random natural processes[38] because the two possible routes are implausible:

1. Either thousands of beneficial mutations would need to occur in one replication cycle, which is statistically impossible (definition of statistical impossibility is clearly defined in the numbers section, and the odds of this far exceed that number), or

2. A sequential series of progeny acquire an increasing number of beneficial mutations over thousands of replication cycles. However, since these are only steps towards a cell wall, and not an actual cell wall, we are still in an open environment and all the bits would float apart. The steps would not be beneficial, and therefore break a key evolutionary principle or law. In the case of a closed environment, there would be insufficient resources available to complete a sufficient number of cycles.

It would appear that we have a chicken and egg paradox that scientific theory, and specifically evolutionary theory, applied to primitive chemical

---

38   For those who have a dangerous amount of little knowledge, horizontal gene transfer in this situation just moves the question to the cell that the code came from – a bit like the meteorite theory.

systems, is unable to solve. This is supported by the fact that there are no viable theories on how a cell wall came into being, just batty compartmentalization nonsense, and yet without the appearance of a cell wall, life could not get more than a step past the starting line.

2.6 SUMMARY OF GETTING TO THE STARTING LINE

So it's time to assess how the things we've learned weighs into our trial of the establishment. Does the sum of knowledge on generating the vital starting materials of DNA and proteins (namely, nucleosides and amino acids) support the case for the defense or the case for the prosecution? And what about the other two issues? How does our knowledge of chirality and the appearance of a cell wall help us determine innocence or guilt? We will assign an arbitrary maximum of 2.5 points that can be gained for each box in the table below, from this section.

1. From chemistry, we learned there is no physical evidence that amino acids or nucleosides were ever produced by random processes. However, there have been some theories that might be chemically feasible. They are a bit ropey, but would it be fair for us to give the "evidence for" column +0.5 for effort? However, the evidence against these theories being capable of forming components in sufficient amounts and purities is many times stronger. Would it therefore be reasonable to add 1 to this column?

2. From chirality, we learned that random processes would produce a mixture of L and D components, which presents a very significant barrier against the belief in random processes. However, I am actually going to use this in the next section, so no score will be added here.

3. The appearance of the cell wall is a paradox. There are no viable theories accounting for it, and the evidence against it occurring via random processes is extremely high. Again, do you feel like we would be unduly biased if we added a 1 to the against column?

4. Since we will not invoke ID, there is no evidence either way regarding intelligence at this point.

So this is the scoreboard at this stage. It's not a good start for the establishment, considering this was about getting to the starting line. Still, there's plenty of time for things to change:

	RANDOM PROCESSES		INTELLIGENT INITIATION	
EVIDENCE	FOR RANDOM PROCESSES	AGAINST RANDOM PROCESSES	FOR INTELLIGENT INITIATION	AGAINST INTELLIGENT INITIATION
AMOUNT (0-10)	0.5	2	0	0

3. THE NUMBERS PROBLEM: THEY ARE REALLY BIG, BUT ARE THEY TOO BIG?

3.1 WHAT IS THE NUMBERS PROBLEM?

Getting to the starting line is getting to the point where you have everything in place to solve the really big question, the one that lies at the heart of life. Although the previous chapter suggested that the chances of even getting that far are exceptionally low to non-existent, for the purposes of being able to focus on this big question, we will work on the assumption that the chemistry and cell wall problems have been solved. As I mentioned, we will carry our findings from analyzing the chirality problem over into this part of our assessment of the evidence.

So what is the really big question we need to answer to determine whether the establishment is guilty of polluting the truth or not?

How did the DNA molecule, the code lying in it, its translating machinery and its replicating machinery all come into existence?

This may seem like many questions, but the different elements are so interconnected and interdependent that, in fact, it needs to be treated as one big question. If there is reasonable evidence that it could have happened by random causes, or if there are viable theories[39] that show it could have, then the establishment are off the hook. However, if there is no evidence supporting random causes, or no viable theories, then it may not be so stupid to believe in the possibility of other causes, and the establishment are guilty.

Remember the criteria for viability from the section "What does evidence-based mean?". In particular, recall that for a theory to be viable, it must be relatively complete, or make **conceptual sense**. Here is a hypothetical[40] theory for the appearance of life from A to L. (Some of these ideas will be explained and discussed later in the text or the appendix):

---

39   There are a number of theories in existence. The few that persist have the appearance of viability because they weren't immediately discredited by the scientific establishment. However, just because they weren't cast out by an establishment desperate to prove a materialistic belief in life's origins, does not mean they necessarily deserve to be protected or nurtured. We will look at the data as objectively as possible to determine whether the word viable or even potentially viable should be stamped on these theories. In the main text, I will summarize briefly the important ones and whether they are viable, but I will reserve the detailed discussions to the appendix to make the main text as accessible as possible.

40   In truth it is not far from the path that a significant proportion of materialistic scientists have faith in.

A. plentiful supply of pure Biological components (assumed)

⬇

B. Appearance of first self-replicating PNA molecule

⬇

C. Transition from PNA to RNA self-replicating molecule

⬇

D. Appearance of cell-wall (assumed)

⬇

E. Generation of self-catalyzing energy source (assumed)

⬇

F. Additional RNA ribozymes that speed up replication

⬇

G. Ribozyme that produces protein from amino acids

⬇

H. RNA aptamers that bind to amino acids

⬇

I. Protein takeover ribozyme roles

⬇

J. Aptamer 'condensation' / merging with co-evovled code

⬇

K. Formation of RNA triplet condon coding system

⬇

L. DNA takeover of code = LUCA - Life as we know it

For this hypothetical proposed pathway to the appearance of life as we know it to be viable, the majority of steps have to be scientifically, statistically and conceptually plausible. There can be no scientific versions of Genesis 1:3 "And God said, 'Let there be light,' and there was light." There can be no "And Dawkins said, let there be an RNA self-replicating molecule, and there was an RNA self-replicating molecule." Unsubstantiated BS will be exposed and cast aside.

The rest of this section will focus on the statistical problems that need to be addressed to find a solution.

Here are some of the theories that have been floating around for some time now: the metabolism first theory; the protein first theory; the protein takeover theory; the DNA genetic takeover theory; and the one most beloved by the atheistic scientific community, the RNA world theory. The ones that still have some sort of traction in the debate, and the ones that are most commonly cited, are covered in the appendix[41]. Whatever theory you pick though, at some point they all have to account for the appearance of large, functional biomolecules, whether they be proteins, DNA, RNA etc. The precise route by which they appear may be a moot point though, since whatever the conceptual merits of a proposed route, all require addressing a central issue: the numbers problem.

The numbers problem is due to the vast statistical odds stacked against the formation of functional biomolecules by a random, undirected process. These odds can be calculated using our extensive knowledge of biochemical systems and put into perspective against the most generous timelines and circumstances. Furthermore, these odds can be measured against widely accepted criteria for what is considered statistically possible.

These calculations, the results of which I will summarize briefly here (and with a bit more detail in the appendix), relate to the cake-baking step, in other words, the assembly of proteins or DNA from a plentiful supply of pure biological components (hence the need to put aside our findings from the previous chapter for now). But first, a thought on how we approach this problem.

---

41   Appendix: Section 3: Theories Other Than the RNA World

3.2 WE ARE WHERE WE ARE, AND WE HAD TO GET HERE SOMEHOW

This sounds like an obvious thing to say, but as I have puzzled over this for more than a decade; it is something that I believe is very important to fully absorb. We are at a destination, a point in the history of life. By default, there must have been a particular path that got us here.

We know precisely where here is. All life on Earth today is based on the same system[42]: the same code, the same translation machinery and the same replication machinery. We have excellent knowledge of this system. That is where we are.

Importantly, if you accept the claims of geologists and paleontologists on this topic, this system has been the only one to exist since very soon after the beginning of life. We know this because, shortly after the Earth was first able to support life, life appeared and then split into three cellular domains, or branches – archaea, bacteria and eukaryotes, which are differentiated by slightly different cellular structures. This is called the tree of life:

---

42   There are a few bacteria with non-canonical codes (reassignments or additions). This is discussed in greater detail in the appendix section 4. In summary, the evidence suggests that these systems existed pre-LUCA and, more importantly, they use the same machinery. It does not point towards the evolution of an additional amino acid.

Phylogenic Tree Of Life

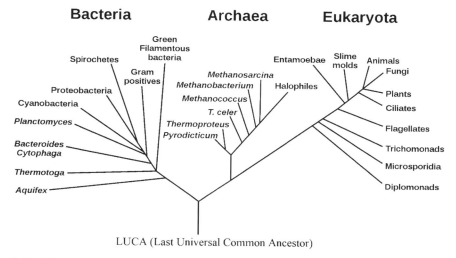

LUCA (Last Universal Common Ancestor)

Original Image created by NASA in 2006 and available on wikipedia. Adapted by Orson Wedgwood

All the different branches on the tree share the same common ancestor, and all have the same genetic system, which is relatively unchanged in all three ever since. This tells us that the fundamental genetic functionality of the different lineages hasn't evolved or changed significantly. Ever. We have been "here," at the destination of the current DNA code and translation system, right from the start of life on Earth. There is no evidence at all that any other code or system ever existed. This is what Francis Crick termed the frozen accident, which I will discuss in more detail later. This supports the idea that life only appeared once on Earth, and that it was always based on the current DNA/protein system.

There are two possible types of route to arrive at this destination. One is via a stepwise series of natural random processes, such as those proposed in the hypothetical pathway I created, and the other is by the spontaneous appearance of the system, *de novo*[43], as it is, in one go.

---

43   Wikipedia - In general usage, de novo (literally 'of new') is a Latin expression used in English to mean 'from the beginning'

There are no other choices available. It was either stepwise, or all in one. No one disagrees with that. Well maybe a relativist might, one who practices spaghetti reasoning rather than logical reasoning.

However, no serious scientist would ever say that a nanoscale system that makes our most advanced technology look like Play-Doh could materialize *de novo* by a random natural process. This means that, by default, if it did in fact appear de novo, it was created by non-natural, non-random processes. However, since the unofficial, modern scientific dogma of only allowing natural explanations for materialistic phenomena discounts this possibility, scientists have no choice but to state religiously that a series of random processes generated this system, even if the evidence doesn't support that position. However, any theories they generate to support this assertion still need to be scientifically, statistically and conceptually sound, and with that in mind, this next point is very important to understand:

Since the current system consists of long chains of amino acids (proteins) and nucleosides (DNA), they had to form at some point. As a result of this fact, no matter what natural route you choose to get to this point, at the very minimum, you need to overcome the statistical barriers described in the next pages.

This is central to understanding why the numbers problem is so important. Whether you have a precursor system which builds gradually (like the metabolism first system), or a system that forms chains randomly from an (imaginary) unlimited supply of biological components, the numbers quoted in the coming discussion are **the smallest possible statistical hurdle**. Why?

The only proposed theory for a natural route to our current system that didn't directly invoke the generation of either proteins first, or DNA (RNA) first, and that stuck around far longer than it should have, is the somewhat discredited[44] metabolism first theory. This is (was) a broad strokes theory that proposes that chemical systems started to develop around geothermal vents due to the excess of

---

44   Vera Vasasa, Eörs Szathmáry and Mauro Santosa. Lack of evolvability in self-sustaining autocatalytic networks: A constraint on the metabolism-first path to the origin of life. PNAS, January 4, 2010

energy coming from these vents. Eventually after "evolving" into self-catalyzing energy systems, they were able to break away and, in turn, sustain the generation of more complex molecules like our biological components. These then began to combine to form biological chains like DNA, RNA, and proteins. If the rest is a bit murky, can you at least see from this last sentence what I was getting at when I said all theories have to pass through the same point?

This theory still requires the generation of functional chains of amino acids - proteins, or nucleosides - DNA (RNA) - the molecules at the core of life's existence[45]. No complete theory is able to avoid the appearance of proteins or DNA in some manner, so for any theory to be viable, it must therefore overcome the numbers problems...it is a mountain that every theory must climb to win. But is it insurmountable?

Even the simplest biologically functional molecule that would have qualified to help life get going would be many hundreds of biological components (amino acids or nucleosides) long[46]. So whether you are considering a protein-first world, a DNA-first world (which would be inactive, so is never considered), a hybrid world, or a precursor world like the RNA-first world (discussed later – for now just think of it as being like DNA)...whatever system you choose, a theory that gets us to the real starting point of life *must* involve the creation of the first biologically functional molecule. This is unavoidable; for life to come into existence, at some point in time this must happen.

So why do these *biologically functional* molecules, which are always linear chains, need to have so many components. Or put another way, why do they need to be so long?

---

45   The metabolism first theory was an attempt to overcome a problem that I didn't even mention in "Getting to the Starting Line" – the development of energy sources and processing fundamental to the continuous running of biochemical systems. The ATP-ADP energy cycle is now that intra-cellular source. Its origins are far from clear, too, but I will not be discussing this.

46   I don't include small polypeptides like cytokines, as these would not be functional in the context of being a vital component of the first-ever self-replicating system.

3.2.1 KNOWLEDGE UPGRADE: WHAT IS BIOLOGICAL FUNCTIONALITY?

Here we need to review some simple biochemistry.

Firstly, what is a *biologically functional molecule*? They are molecules that perform functions beyond just their basic chemical properties and have a specific purpose relating to the production, maintenance and replication of a living system. Recall that a living system is one that uses chemical resources to survive and make copies of itself.

Bacteria are among the simplest living systems[47]. It is a single cell. Some bacteria can remain dormant for ages, but at some point, they come into contact with material (chemicals) that causes them to "wake up" and start consuming that material, which then provides the bacterial cell with the resources to make copies of itself and repeat the process.

If you think about it, humans are just giant, complicated multi-cellular versions of this. Arguably, the most complicated structure in a human is the cell, just like bacteria.

All of the processes involved in making life either arise from, or are mediated by, biologically functional molecules. By now you are familiar with the two major types of functional molecule essential to the emergence of life that we will cover:

1. Molecules that store or carry genetic information: DNA (and, as you will learn, RNA) or;
2. Molecules that mediate or participate in all living processes: this group is almost always proteins, and we will focus on these for the moment.

---

47   Viruses are simpler still but are not considered "living" as they need to hijack the equipment of other cells to replicate.

Proteins are in fact tiny, highly advanced, biological nano-machines that work tirelessly inside us. From processing food and converting it into vital components or energy, to converting this energy into motion, to overseeing the replication and translation of DNA, to building new cells, to laying down bone...this is all performed by or mediated by thousands upon thousands of different types of protein. Proteins are what build and maintain life. (DNA provides the instructions, but more on that later.)

3.2.2 KNOWLEDGE UPGRADE: HOW DO PROTEINS ACHIEVE BIOLOGICAL FUNCTIONALITY?

Proteins can perform a number of different functions (e.g. Antibodies are proteins) but often act like tiny biological machines. The ability of these proteins to perform their functions is entirely due to the *synergistic sum of the protein's chemical properties*. In other words, the role that any protein plays, no matter how complex, is due to all the different basic chemical properties of the individual components of the chain (the amino acids, in this case), working synergistically together to create a specific function.

What is synergy? Synergy is *the interaction or cooperation of two or more organizations, substances, or other agents to produce a combined effect greater than the sum of their separate effects.*

This is precisely what is going with proteins.

As you will shortly see, the advanced function of proteins is due to the way the component parts -- amino acid molecules--work together when linked in a chain. By combining their various chemical properties *synergistically*, you produce a biological machine capable of performing very precise and specific functions which the parts would not be able to perform if they weren't linked together.

It's time to move on from the Victoria sponge analogy and, instead, think of the combustion engine. The various bits and pieces (nuts, bolts, metal shapes, spark plugs, etc.) on their own have limited functions, but once combined (welded, screwed together, etc.), they convert energy (gas) into mechanical power.

A large group of amino acids collected together in the same space, but not joined together in one single chain with a specific sequence, would not get close to performing the function of a protein. Just like nuts and bolts and pieces of metal loosely collected in a basket, they would do nothing. Even if some of the amino acids had joined together in smaller chains, they would not produce the function of the single chain, just as half building the engine would not produce mechanical power. It is only when they all come together as a chain in a precise sequence that function appears.

Since random processes and natural laws do not have purpose or intention, then the appearance of such a molecular machine by random processes would be a complete fluke, or put another way, the result of chance. If this complete fluke is shown to be statistically impossible, then random processes are unlikely to be the cause.

If you look at the stock of available amino acids like an inventory of parts and tools, then common sense suggests that it requires intelligence to put them together in a sequence that generates function. This, however, is only one step removed from the "God of the gaps" argument[48], so I will not invoke it (just as I won't allow "science of the gaps" arguments that rely on a multiverse or quantum mechanics).

---

48   The God of the gaps argument is invoking a supernatural intervention by God for everything that is hard to explain. The Science of the gaps argument is invoking the "science will find an answer" one day for the same purpose.

3.2.3 KNOWLEDGE UPGRADE: PROPERTIES OF AMINO ACIDS THAT SHAPE BIOLOGICAL FUNCTIONALITY

So given that it is the synergy of different amino acids' chemical properties when joined together that give proteins their function, what are the basic chemical properties of amino acids?

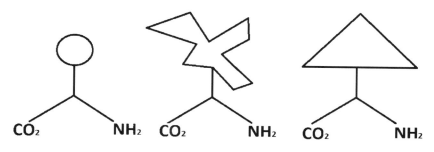

CO_2 NH_2 CO_2 NH_2 CO_2 NH_2

Figure 13

1. Physical structure, or space-taking elements of the structure (bulk). In figure 13, you have a representation of what I mean. Here are three "amino acids." On their own, they have their own properties and their bulk will affect how they interact with other chemicals. They generally avoid each other if possible. When put together in a chain, they are forced into close proximity, but they still want to get into as much space as possible. A property of many bonds is that they are able to rotate, and so if joined together, the groups within these three amino acids would rotate to create the most "comfortable" position. The first position in figure 14 is uncomfortable, a bit like two large people sitting next to each other in economy seats, so this mini-protein can readjust positions so that the big groups are further apart:

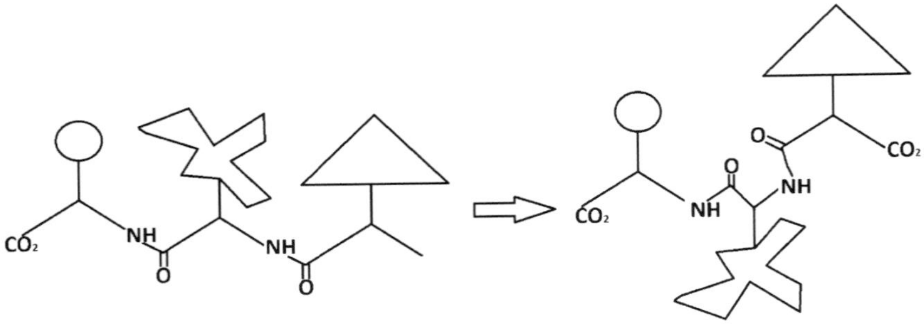

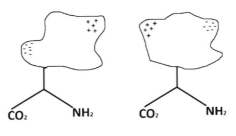

Figure 14

2. "Electrostatic" elements of an amino acid's structure are parts of the structure that have some electrical charge, either negative or positive. This charge causes those parts of the structure to attract or repel other structures that are charged. The "blobs" in Figure 15 represent amino acid side chains with different charged portions. (For a more accurate understanding of the different side chains of amino acids look at the table in the appendix.)

Figure 15

When you join them together in a chain, once again they are repelled or attracted to each other and the structure of the chain can move around to make everyone comfortable:

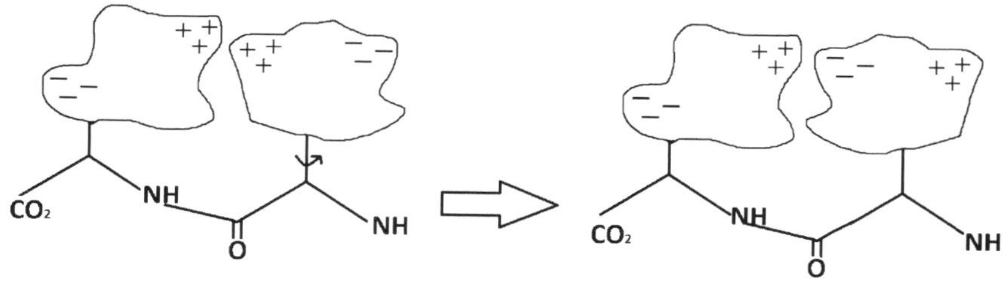

Figure 16

3. Then there are parts of the structure that can actually perform chemical reactions on other molecules. These can form or break bonds and add certain atoms.

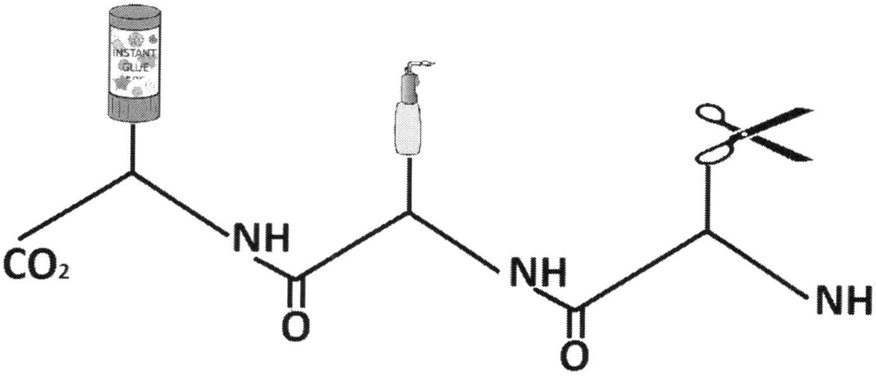

Figure 17

In general, the three types of chemical properties that are contained in one of these biologically functional molecules are physical, electrical and reactive. The *synergistic* sum of these within a chain that contains many hundreds of amino acids produces the required shape and biological function.

3.2.4 KNOWLEDGE UPGRADE: BRINGING IT ALL TOGETHER

Here is actin again[49]. The squiggly shape is generated by all these interactions working together to create the most "comfortable" position for the chain. If you look at the bit highlighted by a box, you may just notice a separate chemical suspended inside the structure. This is ADP (Adenosine diphosphate), which is part of the process that releases energy. In actin, this energy is transformed into mechanical energy in the muscle. Actin (working with myosin) are just like combustion engines; they convert energy into action.

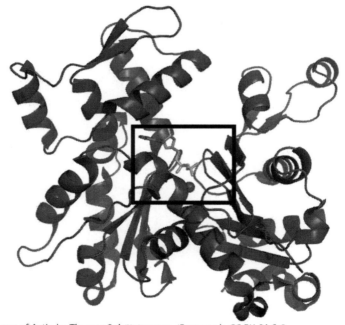

Image of Actin by Thomas Splettstoesser - Own work, CC BY-SA 3.0, https://commons.wikimedia.org/w/index.php?curid=590649. Rectangle added by Orson Wedgwood.

---

49   Picture made available by Wikipedia and created byDeLano, W.L. using the PyMOL Molecular Graphics System (2002), DeLano Scientific, San Carlos CA

The ADP is held in this precise spot because of a specific sequence of amino acids in the actin protein. Any change to this sequence could result in the site no longer being able to hold the ADP and react with it later. This is why you need so many amino acids...precision engineering.

So while individual biological components can be reactive and have unique properties, it is not until many hundreds of them are brought together that they can perform even the simplest functions necessary for living. Around 150 is deemed to be the absolute minimum number required to generate a molecule capable of *doing something meaningful.*

Proteins are the best molecules for performing biological tasks that we know of. There are other types, and I will allude to one significant one later when I discuss the RNA world, however, while you can create chains of other types of chemical that are capable of performing functions, proteins are far and away the champions in the world of nanoscale biological machines. They are quite simply incredible, working at lightning fast speeds, with mind-boggling accuracy. Just to move your arm, hundreds of thousands of proteins are doing stuff...in coordination...in nanoseconds. Mind-boggling.

One of the main reasons they are the best is that there is a sufficient variety of chemical properties in the available component amino acids from which to build a fully functional protein. In general, there are twenty naturally occurring amino acids[50]. Each has different properties, and "levels" of these properties, which are derived from having different side chains. Some are small, some medium-sized, some large. Some are more "charged" or "polarized[51]" than others.

So now that we understand why proteins need at least 150 amino acids to be functional, we are able to actually consider the numbers problem. One final reminder of a central point that underlines the necessity of this section:

---

50   Appendix Section 2. An extra two are covered in the coding section of the appendix.

51   The distribution of charge – one part very negative, another very positive.

any theory that proposes how life came into being must account for the appearances of proteins since they exist and are central to life's existence. Even the much-vaunted RNA world theory (explained and discussed later) must at some point invent or sequester proteins, otherwise it will never become life as we know it. To invent proteins would initially require producing random sequences of amino acids[52]. Likewise, if such a world were to sequester existing proteins, these proteins would also be random sequences. The same applies for any theory, and therefore the following statistical mountain is completely unavoidable.

3.3 THE PROTEIN FIRST, LAST OR ANY POINT BETWEEN THEORY

So, given that proteins are where we are when it comes to biologically functionally molecules, and that they needed to appear for life to begin, we will now consider the statistical hurdle associated with producing a functional protein by random natural processes. We will throw our establishment defendants a bone the size of a T-Rex thigh bone and ignore the fact that even the simplest bacteria requires hundreds of proteins to work in concert to live. For now, we will just consider the chance of one solitary protein forming.

Since a protein needs to be at least 150 amino acids long before it could be considered to have sufficient functionality to be a part of starting life, and because a protein is the most biologically functional molecule that we know of, then this is the smallest protein that should be considered. You couldn't have

---

52   This is the case because the RNA world, or any similar such world, requires the appearance of proteins before a code. (The belief in the appearance of a code for something that doesn't yet exist by random processes requires a level of insanity that has not yet been defined. Bonkers doesn't even come close). Since the sequences aren't coded for, or predetermined, they are, by default, random. This is discussed in detail in the chicken and egg section and in the appendix. It is nitric acid for all these theories.

had bits of protein floating around, or smaller precursors, as they would not have been capable of biological function. Going back to the engine analogy, if you had screwed all the bits and pieces together and now had three large parts of a mostly assembled engine, and that all was required was to join those bits together, the engine would still not work until that had occurred. It would be just as useless as if none of the small component parts had been joined together at all.

3.4 WHAT IS POSSIBLE AND WHAT IS IMPOSSIBLE?

Some argue that if this universe is infinite, then there are infinite possibilities of anything occurring, therefore nothing is impossible. If the universe is infinite, or you have an infinite array of multiverses, then it is certain that there are an infinite number of planets just like ours, and that on one of these planets a cow jumped over the moon[53]. To believe a cow jumped over a moon would require absolute faith in infinite possibilities. That is extreme and irrational faith in a multiverse, possibly to avoid believing something else. It is not evidence-based belief. So do you require faith in a multiverse to believe that a protein containing 150 amino acids could have appeared by random natural processes?

Firstly, as stated, the simplest functional proteins are usually at least 150 amino acids long, and each of those 150 positions has 20 possible options as there are 20 amino acids to choose from.

So if you had a limitless ocean of amino acids and were to randomly start spitting out proteins that were 150 amino acids long, there are 20^{150} (20 multiplied by 20, 150 times) possible combinations that could be produced. In

---

53   No, I do not have a cow fetish.

scientific notation, 20^{150} would actually be written as 1.43×10^{195}. This is a big number. 10^{195} is 1 with 195 zeros after it. To give you an idea of just how big that is, 1 million has 6.

That's a lot of possible proteins, however, the vast majority would in fact be complete junk, as they wouldn't fold properly, and they wouldn't be capable of performing any task. There would be no "synergistic benefit" in the chain of amino acids.

The percentage of proteins that would be junk out of a random sequence of proteins has actually been calculated, and the data published in a formal scientific journal. This number is used in the overall calculation to generate the probability, or chance, of a useful protein forming through random processes[54]. The result of the calculation is coming up. But there are more issues that must be factored in to make this calculation.

In addition to most chains of amino acids that have formed peptide bonds[55] being functionally useless, amino acids have more than one way to react with each other, and so not all reactions would result in these peptide bonds. Every time you add a new amino acid to a growing chain you need to factor this in…i.e there is more than one possible outcome, but only one works.

Now we are going to play our chirality card--namely, that amino acids need to be L-amino acids, and not D. In this calculation, the basic assumption is that there is a 50:50 mix of L or D amino acids in the pool of pure amino acids, and that each time you add a new amino acid to the growing chain, you could have either. However, only one will work. Remember, this is like flipping a coin and expecting heads 150 times in a row. If you didn't try before, have a go now.

Given all these variables, if random processes were generating proteins

---

54   Douglas D. Axe, "Estimating the prevalence of protein sequences adopting functional enzyme folds," Journal of Molecular Biology 2004 Aug 27;341(5):1295-315 – discussed in detail in appendix section 5

55   The unique bond between two amino acids that occurs in a protein chain--the $NH-CO_2H$ bond

from a limitless supply of perfectly pure starting materials, it is estimated that they would have to produce 1×10^{164} (1 with 164 zeros after it) proteins in order for one to be functional. Put another way, if you combine 150 amino acids randomly from a racemic[56] mix, you have a 1 in 1000000000000000000000000 000000000000000000000000000000000000 000000000000000000000000 0000000000000000000000000000000000000000000000000000000000000000 0000000000000 chance of producing a functional protein[57]. This is expressed in statistical terms as a probability of 10^{-164}. A negative exponent (number in superscript) has zeros going to the right of a decimal point i.e. 0.000000...01 with 164 zeros.

Rather than shadow boxing a multiverse[58], let's ask the question of whether this was remotely possible in our understanding of the known universe. That is something that has been quantified, and therefore a question we can answer.

First of all, we need to get a correct perspective of the size of these numbers.

As mentioned, each time you add a zero, you are multiplying that number by 10. Let's see what happens to numbers when you start adding zeros.

A paperclip weighs about 1g. If you add 3 zeros you get 1000 grams, or a kilogram, the weight of an average bag of flour. Add another two zeros and you get 100kg, the weight of a big man.

So with only 5 zeros added, you go from a paperclip to a man.

Let's step it up a bit.

---

56   Mixture of L and D amino acids

57   This number comes from Stephen Myers's book *Signature in the Cell*. For a full explanation of how this number was calculated and the relevant citations, go to Appendix Section 5. There are many other attempts to generate an estimate of the chances of protein coming into existence by chance, all with similarly huge odds against them.

58   Before atheists retreated to the multiverse, they used to say that our universe is so old, and so vast that we cannot comprehend its magnitude and that although the numbers are big, it could happen in our universe. Their need for retreat into the ever more exotic or bizarre, and completely unprovable explanations, is partly due to the conclusions drawn in this chapter by numerous scientists.

If you add 27 zeros, you have the weight of the planet Earth. So by only adding 27 zeros, you go from a paperclip to the planet Earth. We can start to picture just how big these numbers are.

How many zeros do you need to get the estimated mass of the known universe? 56. That's right, by adding 56 zeros onto the end of the weight of the paperclip, you get the mass of the universe in grams. Add one more zero and you get the mass of ten universes. And so on.

So the chances of an event occurring with the odds of 1 in $1X10^{164}$ are ludicrously low, but some would still argue that it is possible.

So at what point do you decide that something is impossible?

There is an arbitrary number of $1X10^{50}$ that is sometimes used, but I prefer the approach that Stephen Meyer used in his book, *Signature in the Cell*. His number is a calculation of the chance of chance happening in the known universe.

The chance, or probability, of a chance (random) event occurring is related to the "probabilistic resources"[59] available. Let's think about a simple, lottery example.

The chance of winning the Wedgwood lottery is one in a million. There are two kinds of chance related to the winning of the lottery.

1. There is the chance that the lottery will be won by an individual ticket, which is one in a million. This is fixed.
2. There is the chance that the lottery will be won at all i.e. whether it will be a rollover. This is variable and related to the probabilistic resources available, which in this case is the total number of tickets that have been sold.

---

59   The probabilistic resources available are the sum of the number of possible *opportunities* that existed for an event to occur. If I buy a thousand lottery tickets, my probabilistic resources for winning the lottery are 1000.

If 750,000 lottery tickets are sold, then the probabilistic resources are high. The chance of chance occurring (in this case, the chance that the lottery will be won) is 750,000 divided by 1,000,000, which is 0.75. Therefore, it is greater than 1 in 2, or a 50% chance that someone will win the lottery, and therefore quite likely. The closer that the chance of chance happening is to 100%, the greater the certainty that the event will occur, in this case that the lottery will be won.

Conversely if only ten tickets are sold, the probabilistic resources are very low (10). The chance of chance occurring is 10 divided by 1,000,000. Therefore, there is a 1 in 100,000 or 0.0001% chance the lottery will be won at all. It is extremely unlikely and would normally be discounted.

Some atheists assert that the universe is so old and huge that, even though the chances of life appearing randomly are incredibly low, the "probabilistic resources" are high, and therefore the chance of it happening is high too. Are they right?

So what are the maximum *imaginable* (not *possible*) probabilistic resources available for a functional protein to form by natural random chances? The maximum imaginable would be if the entire universe consisted of nothing but amino acids, readily available to react, and had been forming proteins at the fastest possible rate since the formation of the universe.

- There are estimated to be about 1×10^{80} atoms in the known universe
- Some evidence suggests the universe has existed for 13.5 billion years
- The smallest unit of time is a jiffy, which is 3×10^{-24} (This means there are 3×10^{24} jiffies per second, yes, 3 with 24 zeros after it. A shedload of jiffies per second).

MIT computer scientist Seth Lloyd calculated that the most "bit" operations the universe could have performed in its history (assuming the entire universe were given over to one single-minded task) is 10^{120} events, or operations. This was calculated by multiplying the number of atoms in the universe, by the time in jiffies that the universe is believed to have existed (by some people). In other words, it is the number of events that could have happened if every atom did something every jiffy since the beginning of time. This is the maximum amount of imaginable probabilistic resources available if the entire universe was devoted to making proteins at a rate faster than chemically possible, where all atoms are, in fact, readily available amino acids. I know--totally potty, but let's roll with it.

Using this number of imaginable events as our probabilistic resource, we are in a position to calculate the chance, or probability, of a protein forming if the universe had done nothing else but make proteins since its beginning, and if the available pool of amino acids was, in fact, all the matter in the universe. That is why this is an imaginary scenario, and one that is extremely generous to the establishment's position.

To calculate the probability of a chance event occurring, we multiply the number of times it had the opportunity to occur by the probability of it occurring once. Let's look at a dice quickly to understand this.

The probability of getting a six after rolling a dice twice is:

2 (the number of times the event had the opportunity to occur) X 1/6 or 0.1666... (the probability of the event occurring once) = 2/6 = 1/3 = 0.333...

Let's now apply this to our 150 amino acid-long protein. Reminder of the numbers:

- Imaginary probabilistic resources of the known universe = 1×10^{120}
- Probability of a functional protein being produced by random processes if only one 150 amino acid protein is produced = 1×10^{-164}

The chance of a functional protein being produced in our imaginary universe is therefore:

$10^{120} \times 10^{-164} = 10^{120-164} = 10^{-44}$ or $1/10^{44}$ or 1 in 10^{44}.

This number is stupidly small 0.000...1 with 43 zeros in total.

Put another way, you would need 10^{44} universes like ours totally dedicated to producing proteins since the beginning of time, and nothing else, to produce a single functional protein. Stuff that in your anthropic[60] pipe and smoke it, Mr. Dawkins.

Poor old Dawkins. Of course, one day he is going to die, and some may feel guilty about saying unpleasant things about him. However, if he is wrong, not only has he insulted billions of people by calling them deluded, but he has also led a significant number of people, including children, away from potentially discovering a very important truth. If he is right, once he is dead, it won't matter if people were rude about him, it's all completely irrelevant. Given his rudeness to so many, I say he is fair game for a bit of ribbing. Anyway, at the time of writing this book, he is still going strong.

So even though the probabilistic resources seem high in our imaginary scenario, they are in fact very low relative to the size of the problem, and still only a chance of one in trillions of trillions of trillions that a protein would form. A statistician, if forced to make a decision, would say that this event is impossible.

Of course, this is a totally unrealistic imaginary universe, but was imagined by some creative post-graduates at MIT to try to put the numbers in perspective. In the real universe, the chance of any protein of even a dozen amino acids long forming by random processes is close to zero, due to the chemistry and lack of available resources.

What are your thoughts on these numbers? Are you someone who still believes nothing is impossible, like the cow jumping over the moon? Personally, I am of the opinion that the chance of a single functional protein coming into existence by random chance is less probable than two cows simultaneously jumping in

---

60   The extreme anthropic principle would say that, because we observe the protein, and that these are the conditions required to observe the protein, then in fact this is precisely what happened.

opposite directions over the moon and hi-fiving each other as they pass.

As stated and shown earlier, to be viable, any theory relying on random natural processes must climb this mountain…there is no other conceivable way around it. None. This is evidenced by the fact that not one theory shows viably how proteins could come into existence. There are only Genesis 1:3-type statements.

But this is just the beginning of the numbers problem. Just one biologically functional molecule on its own would not be a living system. Others would either have needed to appear via the same impossible random route, or this first biologically functional molecule would need to have been capable of generating new functional molecules by itself, of the same or bigger size. This process would still be random, as no molecule would have the "knowledge" of what to make, so each one it makes would be a random sequence, and the same statistics apply…over and over.

3.5 SUMMARY OF THE NUMBERS PROBLEM

We have looked at a number of questions, and this is the state of play:

- We are where we are – life relies on biologically functional molecules, and always has.
- For life to appear by random processes, biologically functional molecules, including proteins, would have to appear by random processes.
- Functional proteins are long chains, with precise sequences.
- To generate such a sequence by random processes is statistically impossible.

So let's ask some questions so we can add numbers to our table of evidence:

- Is there any physical evidence that biologically functional molecules appeared by random processes? No.

- Do any of the origin of life theories explain viably how a protein came into existence by random processes? I discuss these in the appendix. The answer again is no.

- Is there evidence against the *belief* that biologically functional molecules appeared by random natural processes? The chances of such a molecule appearing randomly are so low as to be considered impossible in the known universe. Would you say it's fair to put 2 in the against column?

- Is there evidence for intelligent initiation? Not from our discussion so far.

- There is no evidence against intelligent initiation as this wasn't examined.

So if we add these outcomes to our table it now looks like this:

	RANDOM PROCESSES		INTELLIGENT INITIATION	
EVIDENCE	FOR RANDOM PROCESSES	AGAINST RANDOM PROCESSES	FOR INTELLIGENT INITIATION	AGAINST INTELLIGENT INITIATION
AMOUNT (0-10)	0.5	4	0	0

4. CHICKEN AND EGG PROBLEM: IT'S OLD BUT IS IT ROTTEN?

4.1 INTRODUCTION

Some atheists are prone to eye rolling at the mention of the chicken and egg problem. It has been used by theist apologists for decades now, and many atheists may feel that the problem has been solved. They may even ask, "Haven't you heard of the RNA world theory?" I would answer, "Yes I have."

In this section, we will learn how DNA codes for proteins, and at the same time how DNA is translated by proteins. We will define the chicken and egg paradox that lies at the heart of all life (hint: look at the previous sentence) and ask whether it is still relevant in light of the RNA world theory or any other theories. Is it still fresh?

4.2 IS DNA REALLY A CODE?

Every textbook written on the subject of DNA describes it as a code, but is it really a code, or is the word code just an analogy? The answer is still disputed by some, and there are well-informed people on both sides of the divide who will argue their case forcefully from a technical perspective. The arguments that I have read against the suggestion that DNA is a code rely almost exclusively on their understanding that the code is not arbitrary[61]. I will come back to that. For now, let's learn what code is.

Dictionary definition of code:

1. A system of words, letters, figures, or other symbols substituted for other words, letters, etc., especially for the purposes of secrecy.
2. Computing, program instructions.

To determine whether DNA is a code, we need to see how similar the DNA "code" is to either or both of these definitions, and also to more rigorous definitions like those created by information theorists. Also, we will investigate whether the assignments of amino acid to DNA codons is arbitrary. (This will be explained in detail later.) Before we do, let's look at examples of human codes.

Codes are sequences of motifs (characters, numbers, sounds, electrical impulses, lights, etc.) that contain specific meaning, or information, only when the codes are translated. To translate a code, you need to use a dictionary specific to that code.

Codes can be created for a variety of reasons, but they always have one central purpose, and that is to communicate information between a source and a receiver. Something else they have in common is that, unless you know the code, the meaning of the sequence of motifs is hidden. In other words, the

---

61   Def: based on random choice or personal whim, rather than any reason or system.

code needs to be translated to have any meaning or fulfill a purpose[62].

Codes aren't always just for communicating between people, or even things, that speak the same language. For example, a very obvious use of code is in computers. There are many different levels of coding in a computer. There are higher level codes such as C+, etc., but these all get recoded into machine code, which is a series of 0s and 1s. In fact, these then get further decoded into operands and electrical impulses that cause the machines to perform operations. The original code used by the high-level programmer is completely unrecognizable to the electronic components. By the time the language has been processed, it does mean something to those components, but those instructions would be unrecognizable to the programmer at that point.

Thus, highly sophisticated code can be used to allow two completely different entities to communicate with each other, and is, in fact, sometimes a prerequisite for that communication to occur.

WEDGIE'S CODE

Two good friends at school, Beaky and Wedgie (both names I was called at school, among others that are unprintable) want to be able to communicate during class. (Cell phones are banned.) If they write their messages in English, and the teacher reads the note, they might get in trouble, so they devise a code that only they will know.

Wedgie, thinking he's the clever one, wants to make it hard, and having just learned about codes in math class, starts thinking about how to create a code.

---

62   In its truest sense, all language is a code. Language is a series of sounds, based on an alphabet, that we use to communicate ideas, thoughts, etc. and that our brains translate. Language is a code by which our conscious beings communicate with each other. The letters and sounds are really an abstraction, in that they don't bear any physical relationship to the objects, thoughts, etc. they relate to; the meaning is only learned by understanding the language or code.

He could use dashes and dots, but this is basically Morse code and the teacher might guess.

He decides to use four random letters as the base of his code: A, C, G and T.

He starts by combining these letters in pairs (doublets) and creating the code as follows:

AA=A, AC=B, AT=C, AG=D, CA=E, CC=F, and so on.

The sequence AGAAAG would translate into what? (answer at the bottom)[63]

Wedgie reaches TT and realizes that he wasn't as clever as he thought he was. Why?

AA	A	CA	E	GA	I	TA	M
AC	B	CC	F	GC	J	TC	N
AG	C	CG	G	GG	K	TG	OOPS
AT	D	CT	H	GT	L	TT	

He only has enough possible combinations of doublets to code for sixteen letters of the alphabet, i.e. each pair has two positions to fill, with a choice of four letters at each position, which allows for 4X4 or 4^2, or sixteen possible combinations. Given that the English alphabet is twenty-six characters long, and he would also like some full stops and maybe numbers, he needs about forty. He could devise a completely new code, but Wedgie is also a bit lazy, so instead he decides to just expand his existing code. There are two ways of doing this:

- He could add more letters, say B, F and Y, this would then give 7X7 or forty-nine combinations, or

- He could use the same four letters he's already chosen, but instead of using a doublet system, he could use a triplet system, where the code is read three letters at a time instead of two.

63   DAD

He decides to go for the latter, as it is less complicated. He now has 4X4X4 possible combinations (four possible letters at each of the three positions) which gives him sixty-four possible combinations. (Wedgie got bored at the number 3 and stopped.)

AGT	A	CAT	I	CAA	Q	ATC	Y
ACG	B	AGC	J	CAC	R	CAG	Z
AGA	C	GAT	K	CTC	S	CTG	.
AAT	D	AAA	L	CGG	T	GAA	,
GAC	E	ATA	M	CCA	U	ATT	?
ACC	F	CTT	N	CGA	V	GTA	1
ATG	G	AAC	O	CTA	W	ACA	2
ACT	H	GCA	P	AAG	X	AGG	3

So using this new code, what does the sequence CACAGTCGG translate into? Remember, you are now reading in sequences of three letters at a time (answer at the bottom)[64].

Did it take you a few seconds to find the first two letters? That is because Wedgie might be lazy, but he isn't stupid, and wanted to ensure any obvious connection between the triplets and their meaning was removed. This is what makes this code truly arbitrary. This is a very important, possibly central, concept to understand if we are going to determine whether DNA is indeed a code. A system like the above is only a genuine code if there is no obvious, or visible, link between the source characters and the meaning. The code is chosen randomly. AGT means A, but could just as easily have meant V, if we had chosen it to. There is no intrinsic link between the sequence AGT and A outside of the code. The assignments are not traceable to the pattern without the translation sheet.

---

64   CAC AGT CGG, which is R A T

DR. WEDGWOOD'S CODE

Wedgie has grown up and is now a senior cryptologist (code specialist) at the CIA. He has been tasked with creating a system by which English-speaking CIA operatives in Langley can communicate with their Arabic-speaking agents in Iran. To do this he must:

- Create a code using English characters which the CIA handler inputs in Langley
- Use a vocabulary of about 1,000 common words
- Create a machine that reads the English characters and translates them into symbols which are received by the agent in Iran

Dr. Wedgwood recalls the halcyon days of his misspent youth, and the code he devised with his friend Beaky, and partly from nostalgia, and partly because he is a bit of a lazy bugger, he decides to use their secret code as the basis for this new code. He uses the same letters A, C, T and G, and he uses triplet combinations, only this time, instead of the triplets coding for letters in the English alphabet, they code for the following Arabic symbols[65].

---

65   If you know Arabic, you will know that each letter has 4 different variants depending on location. Wedgie decided to overcome this by using start and stop to signify the beginning and end of the word. The machine then rearranges everything as Arabic is right to left.

AAA	ط	TAA	ض	TGT	٠ ▲
AAC	ذ	GAA	STOP	TCT	ش
AAT	ص	CAA	غ	ATT	ى
AAG	ف	TTA	ب	GTT	ح
ACA	ج	TTG	خ	CTT	ع
ATA	،	TTC	ق	CCA	ا
AGA	ر	TAT	ز	CCT	ن
ACC	ك	TCC	ث	GCC	د
CAC	ت	CTC	م	CGC	ة
GGC	س	GGA	ط	GGG	START

But there is a problem. Dr. Wedgwood gets an offer from a Silicon Valley company to come and work for them, and he leaves the CIA before he finishes the machine. It only codes one way. The CIA, having invested all this time in the process, decides to use it anyway.

The CIA boss can't speak Arabic and is only given a vocabulary of words in English in the English alphabet code.

E.g. in his vocabulary book "banana" = GGGGTTTATGAA

He types this into his computer, which sends the message to Iran.

The decoding machine in Iran reads the message and translates the sequence into the corresponding Arabic sequence of symbols which is banana, or "muz" in arabic (using the correct version for position etc which this isn't): جز

So this is the essence of the code, and there are a few things to note:

- Without the translation machine, there would be no means by which the English-speaking CIA boss and his spy in Tehran would be able to communicate. If they picked up the phone, they would not understand each other.
- Again, the code is arbitrary, there is nothing intrinsic in the triplet English alphabet codes for each Arabic symbol that could be associated with the assigned Arabic symbol. The assignments,

while generated by intelligence, are random.

- The machine would need to have been made by someone who spoke both English and Arabic; an entity that was outside of the system.

So given that we understand that this latter example is a code and that the assignments are arbitrary, we can compare it to DNA's amino acid "code" and determine whether DNA is a real code and if so we can remove those quote marks.

THE DNA "CODE"

The DNA "code" does not actually consist of Roman alphabet letters, but as mentioned earlier is made up using four nucleosides. However, we do in fact represent the four nucleosides with the letters A, C, T and G when writing the DNA code[66]. That is our representation of them, our code, but if we are to determine whether DNA is a true code, then we need to establish if the nucleosides are arbitrary in relation to the amino acids they code for. Are nucleosides, for all intents and purposes, just symbols as well? This question also lies at the heart of the chicken and egg question.

The structure of DNA was first discovered by James Watson and Francis Crick, in Cambridge University in the U.K., in 1953. This is the famous double helix which everyone now associates with DNA. For those who don't have knowledge of chemistry and biochemistry, it looks very complicated, but in fact it is incredibly elegant and simple, as I showed earlier. As a reminder, the double helix consists of two strands of DNA. Each strand of DNA consists of billions of nucleosides joined together in a linear fashion, i.e. they link together to form one chain.

---

66   Details about the RNA code are in the appendix, but the differences are like speaking English with an English accent or speaking it with an American accent. They are otherwise the same.

Earlier, I used very simple pentagons to represent nucleosides to show what the components of DNA were and how they formed a chain. Below are the complete structures using chemical shorthand. They are called: [deoxy] Adenosine, [deoxy]Cytosine, [deoxy]Thymidine and [deoxy]Guanosine and are given the letter assignments A,G,T and C (the capitals are added to help show where the letters are derived from):

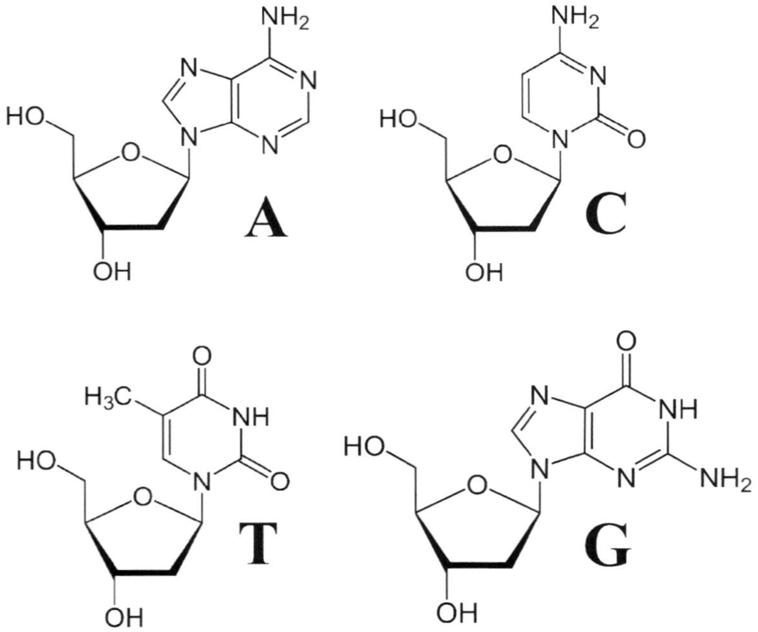

They look quite complicated but, considering that the instructions of every organism on the planet are written using only combinations of these four chemicals, they look embarrassingly simple. This is a remarkable truth about life.

Let it sink in.

Everything about your physical body is "coded" using those four chemicals and nothing else. By writing out a sequence of nucleosides about 3 billion long, using just combinations of those four letters, every organ, every nerve, every muscle, every bone…everything about you is described. But there's more than that. Within that sequence are instructions of when to start building those bits and pieces, how to maintain them, and when they should stop growing. Just four chemicals, combined in a specific way, are capable of communicating vast amounts of information more complex than any human has ever begun to even conceive. Excuse me while my objectivity goes AWOL for a second.

That is fricking *genius.*

I don't care how unscientific it sounds; in my opinion, anyone who thinks for a second that could be the result of natural random processes is a few sandwiches short of a picnic. To me, scientists who don't see this are like fleas on the back of an elephant; they understand everything there is to know about the skin of an elephant and are able to describe its structure and characteristics in perfect detail, but they don't believe in elephants. They've never seen an elephant, even though they've been sitting on one all their lives. They've got so lost in the detail that they can't see the big picture. The origin of DNA is, for me, the scientific elephant in the lab.

Sorry…just needed to get that out of my system.

While this amazing system might obviously look and feel like it is the result of intelligence, that feeling is just intuition. That doesn't mean it is wrong, but that it is subjective. To answer the question of whether the establishment has been lying to us on this central question of the origin of our existence, we need objective evidence that either supports this intuition or, in fact, shows that random processes could have generated it. Solving the chicken and egg riddle in the context of DNA and proteins, i.e. which came first, will help provide the answer to that issue. Also, understanding whether DNA is an arbitrary code will help us to answer the other question about our origins: Is there evidence for intelligence? So let's keep digging into our understanding of DNA and the coding system, and maybe we will be able to answer these questions.

The nucleosides in DNA, A, C, T and G, have one solitary purpose – to convey information. **Each nucleoside in DNA has the sole function of**

representing one single unit of information[67]. For all intents and purposes, it is just a symbol, motif or a letter. You might ask whether this is strictly accurate since this is a chemical; how can a chemical be information? If you are reading this in a book, then the letters on the page before you are chemicals (those comprising ink) conveying information. If you are reading this on an LCD screen you are looking at chemicals (liquid crystals) conveying information. Electrical impulses can convey information. Sounds can convey information, light, and so on. It is not the medium that determines whether it is information; it is whether the chemical is being more than just a chemical. It is whether the chemical has *meaning* beyond its intrinsic properties.

The nucleosides in DNA do have meaning, and we represent that using the letters A, C, T and G which, in reality, is no different from drawing the chemical structures. From now on, I am going to refer to the nucleosides in DNA using letters. That is how scientists write out DNA "code" -- as a sequence of letters. The entire sequence of letters is an instruction manual, and that is a precise description. DNA is a passive molecule; it does nothing. It is read like a book and conveys information in the same way.

So how on Earth can you write the instructions to build and maintain a human, one of the most complex organisms to have ever existed on Earth, and that consist of trillions of different cells, using just four letters?

You already know the answer. The analogous Iranian spy code that I created is composed of the same four letters, and yet, by combining them in triplets, we are able to generate a code for all twenty-eight symbols in the Arabic alphabet, and punctuation. (As a reminder there are sixty-four possible combinations). Given that you can write whatever you want in Arabic, there are no limits to the size or complexity of the messages, except those imposed by the dictionary of a few thousand words. If he had wanted to, he could have included every word ever used in Arabic, but we know that Dr. Wedgwood is not the hardest

---

67   In computer parlance, this is called a bit, and in fact the size of an organism's genome is often expressed in terms of megabytes (MB) or gigabytes (GB).

worker on the planet, so he stopped at a thousand.

The same is true for the DNA "code." Just as with the CIA code, for each triplet there are 4 X 4 X 4 possible combinations of the letters A, C, T and G, i.e. sixty-four possible combinations (AAA, AAC, ACC, ACA...TTG, TGG, GGG).

The DNA in every cell in your body is an enormous string of letters, read in sequence of three letters at a time. The combinations of the three letters, triplets (or **codons,** as they are called) are translated by a biological machine, called the ribosome, with the help of some other molecules that all live in the cell (explained shortly). They perform the same function as Dr. Wedgwood's machine, and this system also suffers from the same issue of only translating in one direction.

DNA, which is composed of nucleosides, "codes" for a completely different and unrelated chemical system, amino acids, in the same way that the Roman letters in my analogy code for Arabic symbols. Each triplet, or codon, "codes" for a single amino acid. As you will recall, amino acids are the building blocks of proteins.

In total, there are twenty naturally occurring amino acids (the two others are covered in the appendix). Therefore, DNA codes for twenty amino acids. It also codes for full stops; quite literally, there are stop codons that tell the machinery inside the cell when a specific sequence has come to an end. Given that there are sixty-four different triplet combinations available, there is "redundancy" built into the system. (I cover this and Junk DNA in the appendix).

Below is an example of a codon and the matching amino acid it codes for. The molecules on the left are read just like letters, in this case A-T-C; this only ever means (is translated into) the amino acid isoleucine (shorthand Ile), which is the structure on the right.

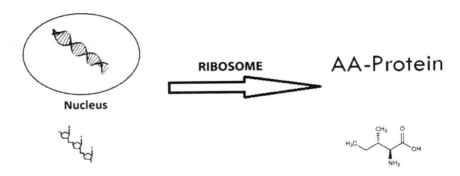

DNA Triplet Codon   Amino Acid

The next diagram is a basic schematic of the overall process. (The exact details of how it happens, and what a ribosome is, will be explained in greater detail in Knowledge Upgrade 4.3.1).

Nucleus   **RIBOSOME**   **AA-Protein**

A segment of DNA is copied and leaves the nucleus in a slightly different form. The ribosome then "reads" this segment of DNA from one end to the other, just like a sentence. As it is doing so, the DNA is translated codon by codon (triplets of 3 letters) into the corresponding amino acid. The ribosome then links these amino acids in a linear chain to create proteins that then fold into specific structures, which perform biological functions.

So are we in a position to answer the question, "Is DNA a code?" Yes.

There is a saying that is useful at this stage. If it walks like a duck, swims like a duck and quacks like a duck, then it probably is a duck. Looking at how DNA functions, namely it is a passive linear sequence of chemical "motifs" composing triplet codons with the sole purpose of representing, or symbolizing, amino acids and the only way it becomes of use, or acquires meaning, is by being read and translated by a bespoke system, then it is fair to say that it looks like a code, reads like a code and is translated like a code, therefore it is a code. That is why it is described in every textbook as a code; there is no other description that is able to more accurately capture its essence, and that in itself is evidence that it meets the criteria of being a code. Let's go back to our dictionary definition and apply it to DNA:

1. A system of ~~words, letters, figures, or other~~ symbols (*amino acids*) substituted for ~~other words, letters, etc.~~, symbols (nucleosides) ~~especially for the purposes of secrecy~~.

2. ~~Computing program~~ instructions *for making a living organism*.

OK it needs some editing, but actually the nucleoside-amino acid code has the characteristics of a "secret" code but acts like a functional code with instructions.

Moreover, Hubert Yockey, who was an international expert on bioinformatics, and someone who had a far superior knowledge of information theory than I ever intend to, wrote a book on the subject called *Information Theory, Evolution and the Origin of Life*. (2005, Cambridge University Press). In this work, he measured all the properties of DNA against the various criteria and principles that we use to define codes, and he showed that DNA meets all these criteria. Anyone who read his blog will know that he was no fan of Intelligent Design, and he didn't invoke an intelligent creator. Instead he used another word to describe the conundrum of the appearance of life, and specifically the DNA code, on Earth. Unsolvable. I will come back to that at the end of this book.

Suffice to say, if someone like Yockey using strict criteria said that DNA walks like a code, swims like a code and quacks like a code, then I am very much inclined to believe it is a code. What about you? Yockey's critics say he was wrong because the code is not arbitrary: the codons have a physical/

chemical connection or relationship to the amino acids they code for, therefore they are more than just symbols. We will discuss this in a moment by looking at the heart of the chicken and egg problem--the central problem facing any random process argument.

However, given what an expert on information theory like Yockey said, I will no longer put quote marks around the word code.

So to summarize the DNA code: It is simple and elegant and yet capable of coding for everything that is alive today and that has ever lived.

- The code for all living things is contained in the DNA of each cell of every being.
- The string of DNA is one enormous linear sequence of nucleosides, or letters, which is read sequentially, just like a book.
- Only four nucleosides, or letters, are used.
- The code is compiled of triplet combinations using these four "letters," creating 64 possible combinations.
- The code is read in a linear fashion and translated sequentially by the ribosome and other molecules to create a corresponding string of amino acids.
- These chains of amino acids are proteins and do all the work of life.

Are you excited? I hope so. You are about to learn about one of the greatest mysteries in life.

4.3 THE RIBOSOME: OFTEN CALLED THE TRANSLATION MACHINE...BUT IS IT?

The ribosome is a biological "machine" that formed at life's beginnings. Using a word like "machine" might cause some to feel that I am inferring Intelligent Design. If I was to ever invoke ID in this book, then this would indeed be the moment, since the ribosome could be seen as Intelligent Design on steroids. Indeed, while we go through this, you may find it hard not to intuitively feel that the ribosome is a machine and was the result of design, but in this instance the word machine is an analogy and we will not count this in the evidence table. Either way, the subject of the ribosome cannot be avoided since it is one of the components involved in translation of DNA, and it is therefore central to understanding how DNA is translated. Calling it a machine may be analogous, but it best sums up what it, and so many other biological constructions, are like, and how they behave.

The ribosome consists of an array of proteins and RNA molecules that work together to read the nucleoside chain. The ribosome, with the help of some additional molecules, some of which are proteins, and some of which are RNA molecules, translates nucleosides into amino acids. These amino acids are joined together to make a protein. The details will follow.

The big question at the center of the origin of life puzzle and the chicken and egg paradox is now right in front of our eyes. Given that proteins are involved in the translation of DNA, and yet these same proteins are coded for by the DNA they are translating, which came first? It appears you can't have one without the other, therefore it appears to be an unsolvable paradox. But is it?

Proponents of the RNA world theory, which we will come to later, say it is solvable and that the presence of RNA in the translation process is a living, biochemical fossil providing evidence of their theory. To assess the validity of this, we need to gain a full and precise understanding of each step in the translation process so that we can determine the extent to which RNA is responsible for the translation of nucleosides to amino acids. If it is entirely down to proteins, then the chicken and egg argument holds water, but if RNA is responsible for the actual translation, then the RNA world theory may get a foot in the door.

4.3.1 KNOWLEDGE UPGRADE: THE TRANSLATION PROCESS

The rather complicated diagram below is a standard high school (or, in this case, adapted from Wikipedia[68]) depiction of how DNA is translated.

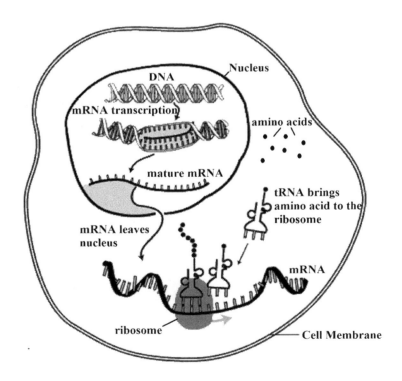

---

68   Source: https://en.wikipedia.org/wiki/Protein_biosynthesis

The overall process shows how we go from the DNA code to a complete protein, but to identify exactly where translation occurs, we need to break it down into sequential steps.

In the first step, called transcription, the DNA double helix is opened up and a chain of RNA nucleosides[69], called mRNA is constructed along it by a protein called DNA polymerase. mRNA (messenger RNA) is just a copy or mobile form of a segment of the DNA code called a gene. DNA is so precious that it is kept in the hallowed environment of the nucleus where it won't get damaged.

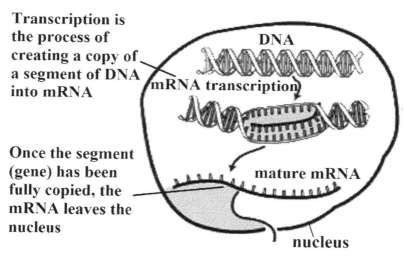

Transcription is the process of creating a copy of a segment of DNA into mRNA

DNA

mRNA transcription

Once the segment (gene) has been fully copied, the mRNA leaves the nucleus

mature mRNA

nucleus

Transcription Process

The nucleus is like a library in Cambridge University or Harvard. The knowledge contained in those books is highly valued and protected. Unlike a university library, which allows you to borrow its books, the nucleus guards the DNA jealously, and will only allow copies to be made for use outside of its walls. If DNA is a book, then mRNA is a copy of a section from that book that describes how to build a protein. At this stage, mRNA is still in the form of a

---

69   Appendix Section 6 – structure of RNA and anticodons etc. are discussed

nucleoside code. **No translation has occurred.**

In the next step - usually described as the translation step - this strand of mRNA leaves the nucleus of the cell and heads for the ribosome, which is a chemical machine that uses the mRNA as a template to construct proteins. The ribosome is comprised of rRNA (ribosomal RNA, which is functional rather than code) and proteins mixed together in a blob.

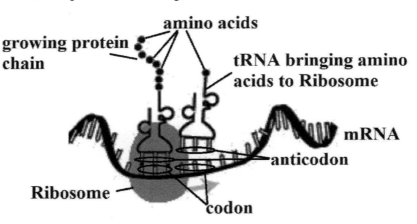

"Translation" of mRNA codons into amino acids

They work together to hold the mRNA in place, grab the correct tRNA (transfer RNA – much more on this in a moment), and bind the new amino acid to the growing chain that will eventually become a protein.

So let's break this step down further to fully understand what is going on, and whether this is indeed where translation occurs. These are the things that happen:

- The mRNA becomes attached to the ribosome and is held in place. Remember, mRNA is in essence just a mobile segment of DNA; it is still in nucleoside code.
- tRNA molecules, which are already attached to a specific amino acid are brought into the ribosome. On one end of the tRNA is an

anticodon, which is a specific sequence of 3 nucleosides that is attracted to a specific codon on the mRNA. The matching codon on the mRNA obviously codes for the amino acid attached to the tRNA.

- Due to this attraction, this anticodon on the tRNA aligns with the matching codon on the mRNA and they "stick" together.

- At the other end of the tRNA molecule, the attached amino acid is brought next to the previous amino acid in the growing polypeptide chain. (A polypeptide is a chain of amino acids; it is only called a protein once it is complete and able to carry out the function it is "designed" for.)

- The amino acid is then bound to the chain and the tRNAs leave to find more amino acids to bring to the ribosome.

- The mRNA is shuffled along and the process repeated until the protein is complete.

4.4 DOES tRNA TRANSLATE NUCLEOSIDES INTO AMINO ACIDS OR ARE KIDS RECEIVING A FAKE EDUCATION?

So from the Wikipedia/high school account of translation, has translation occurred, and was RNA responsible? It certainly seems to have. In high school textbooks, and other simple accounts of the translation process, credit is given to tRNA for being the central molecule responsible for translating the code.

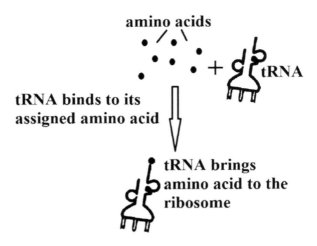

amino acids

tRNA binds to its
assigned amino acid

+ tRNA

tRNA brings
amino acid to the
ribosome

The Real Translation step?

If you look at this part of the original diagram, it appears that tRNA does, indeed, do the translation since, from this common representation, it finds the corresponding amino acid all by itself and binds to it. tRNA is often cited as being the molecule that solves the chicken and egg paradox. Since RNA is very similar to DNA, and since it apparently also translates the DNA, it suggests that proteins were at one point not needed. This is central to the RNA world theory.

But is this true? Does tRNA actually translate DNA? If not, then kids may indeed be receiving a fake education.

To answer this question, you need to ask a more precise question:

Does tRNA itself identify and bind to the amino acids it brings to the ribosome, or is something more going on?

If it does, then the paradox is not a paradox. Why? If tRNA is able to identify amino acids unaided, then it shows a physical/chemical relationship between nucleoside code and amino acids. This would not only show that the code is not arbitrary but would also hint at a precursor system…i.e. it points to a previous step in an evolutionary path from a simpler translation system, aka the RNA world.

So firstly, what is tRNA?

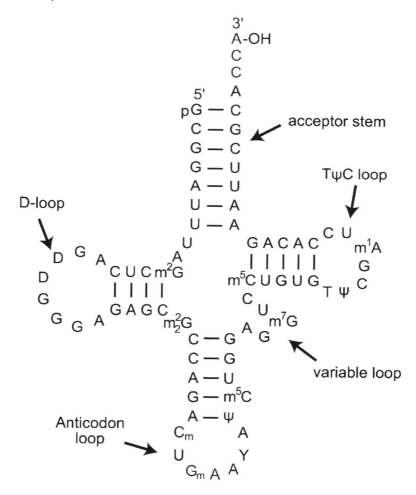

Above is a rather frightening image of a tRNA molecule[70]. As mentioned, tRNA actually stands for *transfer* RNA because it transfers the amino acid

---

70   By Yikrazuul - Own work, CC BY-SA 3.0, https://commons.wikimedia.org/w/index. php?curid=10126790

from one part of the cell to the ribosome. You will notice that it is represented as a sequence of letters, very similar to DNA. The observant might notice that instead of A, C, G and T, in RNA we have A, C, G and U. In DNA the T stands for thymidine and in RNA the U stands for uracil. The difference between the two is small, and they have the same meaning when translated. It's like the difference in spelling between British English and American English, and the swapping of s with z. In British English it's emphasise and in American English it is emphasize. But it has the same meaning. ATC in DNA has the same meaning as AUC in RNA, and if you are still with me, that is the sequence for isoleucine.

There is one tRNA for each codon of the DNA code. (Remember, a codon is the triplet that codes for a specific amino acid.)

If the tRNA does indeed bind to, or grab, the appropriate amino acid all by itself, and transfer it to the ribosome, then the RNA world theory may hold water, as the DNA/protein chicken and egg paradox would be resolved. Why?

It would show that, while proteins help with translation, or speed it up, the central step (namely, the identification or the association of an amino acid with its codon) is performed by RNA. Therefore, DNA does not **need** proteins to be translated. The answer to the question of whether tRNA is the molecule responsible for translation is, therefore, probably the biggest question in biology. You'll be glad to know there is an answer.

There is a very important section of the tRNA that is highlighted in the diagram: a nucleoside triplet called the anticodon loop.

Remember, mRNA was the mobile form of a whole section of the DNA code. tRNA is in essence the mobile form of a single codon, or a triplet, a sequence of three nucleosides that codes for an amino acid (the anticodon might be better described as being similar to a negative of a photograph, but that would be completely lost on anyone below the age of 35)

You'll notice in the diagram of the whole process, by the time the tRNA arrives at the ribosome, it is carrying the assigned amino acid, and it binds to the corresponding mRNA codon. A bit of chemistry goes on at the other end of the molecule, the amino acid that was attached now joins the chain of amino acids, and the protein is formed.

So the key question is, does this anticodon portion of the tRNA molecule also identify and bind to the amino acid it carries to the ribosome?

No, it does not.

Does any other unique part of the tRNA identify the amino acid?

No it does not?

Really?

Yes. Really.

tRNA, whether it be the whole molecule, or a part of it, even the anticodon, never uniquely identifies or grabs the amino acid it is assigned to. If you chuck the tRNA molecule assigned to isoleucine into a bucket of only isoleucine, it still wouldn't find it.

There is nothing in the structure of a tRNA molecule that cause it to directly identify the amino acid that the nucleoside triplets code for. There is no physio-chemical relationship, the relationship is abstract and arbitrary.

This is one of, if not the most important fact in all biochemistry. Going back to my English-Arabic analogy, there is nothing about the English alphabet triplets that code for the Arabic symbols that allows you to associate the two. There are no properties that make it obvious which triplet of letters code for their Arabic symbol…you could look at them all day, and without the translation code, you would not be able to guess because Dr. Wedgwood's code is arbitrary.

This also applies to the tRNA molecules. You might think that, given the more complex structure of the tRNA molecule, that there might be some region that plays a part in directly identifying the amino acid. But there isn't, as you will see in a few moments.

tRNA is not the superstar that some would have it be. Rather than being the Atlas-like wonder molecule upholding the RNA world, it is just a lowly mobile version of a codon, and a passive player in the translation process.

So if it's not tRNA, then what is responsible for the critical translation step? Chicken…egg…DNA… anyone, anyone…I see a hand at the back.

It's Wikipedia, the internet know it all, with its hand up…of course. Keen to redeem itself, Wikipedia points out that on the same protein biosynthesis page (https://en.wikipedia.org/wiki/Protein_biosynthesis) on which the previous diagram was found, there is a more detailed diagram[71] showing the translation step in greater detail. This is shown below in the highlighted step one.

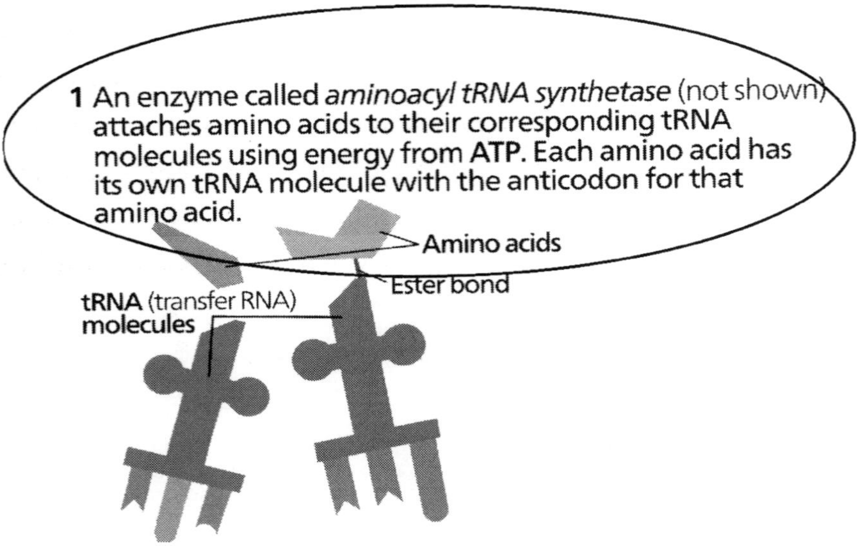

1 An enzyme called *aminoacyl tRNA synthetase* (not shown) attaches amino acids to their corresponding tRNA molecules using energy from **ATP**. Each amino acid has its own tRNA molecule with the anticodon for that amino acid.

Amino acids

Ester bond

tRNA (transfer RNA) molecules

---

71   Created by Kelvinsong: https://commons.wikimedia.org/wiki/File:Protein_synthesis.svg

4.5 ARS

Knowing that I lived in North America and that I am British, you might be wondering if this is some sort of hybrid spelling[72] of an impolite word referring to a part of the anatomy used for sitting, among other things. While I sometimes get my sidewalks and pavements, or my arugula and rocket muddled up, I am not trying to randomly introduce inappropriate language.

ARS is actually the commonly used acronym for the protein amino acyl-tRNA-synthetase. Sometimes aaRS is also used.

Yes, you guessed it, the actual point of translation is, in fact, mediated by a protein--the ARS protein. There are twenty of these proteins, one for each amino acid, and they are molecules that have unique binding sites for both the tRNA and the amino acid that each specific tRNA is associated with[73]. This is the molecule that actually does the translation, and since it is a protein, and is coded for by DNA, it is the molecule that lies at the heart of, and exemplifies, the chicken and egg problem.

This is the process below[74]. Just focus on the key points.

---

72   U.S.- Ass; U.K. arse

73   There are sixty-one tRNAs due to redundancy, but only twenty ARS. However, the twenty ARS can bind to the different tRNAs coding for the same specific amino acid. Yes, it's a head scratcher, but the ARS uses multiple domains on the tRNA to do this, along with the anti-codon section. Crucially, the tRNA never locates the amino acid directly. Refer to Appendix Section 7.

74   Created by Robert Maxwell: http://biomoocnews.blogspot.ca/2012_10_01_archive.html

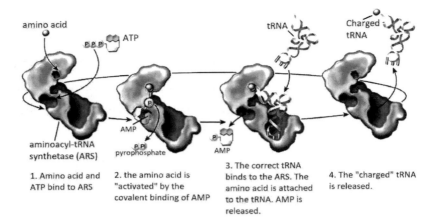

amino acid
ATP
P P P

aminoacyl-tRNA synthetase (ARS)
AMP
P Pi
pyrophosphate
P
AMP

tRNA
Charged tRNA

1. Amino acid and ATP bind to ARS

2. the amino acid is "activated" by the covalent binding of AMP

3. The correct tRNA binds to the ARS. The amino acid is attached to the tRNA. AMP is released.

4. The "charged" tRNA is released.

The blob is the ARS. Blobs are an easier way of representing proteins than the ribbon structures that I have shown before. The ARS protein first seeks out the amino acid that it specifically works with (remember there are twenty different ARS proteins). It then gets fuel for a reaction. (ATP is the gas of life.) Once it is good to go, the ARS finds the right tRNA.

At no point does the tRNA identify the amino acid. The end of the tRNA that binds to the amino acid is identical in all tRNAs, so is not even closely implicated with identifying the amino acid. The ARS brings the amino acid and the tRNA together and it forms the bond between them, using the energy from ATP.

This is the precise step that results in translation. This is the point at which the nucleoside code (originally DNA, then its non-identical twin, RNA) is translated into amino acids. The mobile form of the codon, the tRNA, is passive throughout the process.

Truly, you can say that without proteins, DNA would not get translated, and yet without DNA, these very proteins would not exist. RNA is involved in the process only in so much as it is a more mobile and agile form of DNA, but it does not do the translating, and there is no evidence whatsoever that it ever did. At no point does DNA or RNA ever actually associate with the amino acid it codes for without the mediation of these ARS proteins.

I hope all this is clear. I know that when I come across something new like this, I need to read it through a few times before I fully internalize it. I believe it is worth doing that in this case, so you will really grasp both the simplicity and elegance of the code, and the complexity of the paradox at the heart of its translation. It is the central puzzle of the whole question of how life came into being.

So we can see that the chicken and egg paradox is intact. There is no obvious way as to how random processes could solve this paradox, since it would require the appearance of the translation machinery *de novo*, but that is impossible. Moreover, we have answered another important question.

We have shown that there is no physical or chemical link between the codons or anticodons and the amino acids they code for.[75] The assignment of the codons to the amino acids is, therefore, random and arbitrary[76]. This presents a very significant barrier to the belief that random causes could have generated the code, as there is no conceivable route by which an arbitrary code might be generated in this way--a point that is underlined by the lack of any theory. Also, the fact the code is arbitrary has relevance for the last evidence section.

---

75   Michael Yarus, a biochemist from the University of Colorado, has spent many years suggesting that there is a physical link between codons and their amino acids due to a correlation between codon concentration in binding sites and the amino acid they bind in man-made RNA aptamers. This is discussed in detail in the appendix, but I ultimately conclude it is a red herring and does not undermine the arbitrary nature of codon assignment in the DNA/protein code. There are other scientists who agree.

76   This point is somewhat exemplified by the fact that more than one codon can code for a single amino acid (but not the other way 'round). For example, both TTA and CTG code for leucine. Neither of these codons or the anti-codons directly binds leucine.

4.6 RNA WORLD THEORY

RNA is very, very similar to DNA. Structurally, it is almost identical in that it is a nucleoside, but with two minor changes[77]. It is a bit like the difference between American English and British English, but it has important additional properties, too. Not only is RNA able to store information in the form of code like DNA, but it can also perform protein-like functions. This lends itself to being an ideal candidate for a precursor world, where one molecule is able to do both the functions of DNA and proteins. The belief is that once this precursor RNA world reached a certain level of sophistication, it was replaced by DNA, a more stable form of information storage, and RNA biological machines were replaced by proteins, which are many orders of magnitude more efficient.

This is pretty much how it is described everywhere, with different variations. However, in my opinion, and others', this theory has more holes in it than a sieve. This is a brief summary of objections thrown up by both atheist and theist scientist over the years, which are discussed at length in the appendix:

- RNA is unstable and would not have hung around in the early Earth environment long enough to form long complex chains capable of doing anything. To overcome this, precursor worlds of the RNA world (which is already a precursor world of the real, DNA world) have been suggested, such as the PNA world. Transitioning between these different worlds would be a bit like transitioning between parallel universes.

- The statistical hurdles of creating a single chain of RNA capable of being biologically functional or being able to self-replicate are not dissimilar to those encountered in the big numbers section

---

77   The structure of RNA nucleosides and various functions of RNA molecules are discussed in Appendix Section 6

- As mentioned previously, for proteins to enter the RNA world requires overcoming exactly the same statistical hurdles as those defined in the numbers section...on top of the hurdles required to form the first functional RNA-only world. Moreover, no conceptual route has been proposed to explain why or how RNA would suddenly incorporate proteins as part of its system.

- There is no plausible explanation of how the system transitions from RNA to DNA.

- No one, ever, proposes a viable mechanism by which RNA becomes a code for proteins, or why it would.

So if we were to measure the RNA world theory against the required criteria for defining a viable theory--namely, that there needs to be a reasonably complete sequence of chemically, statistically and conceptually viable steps--it is an epic failure. If we are going from A to LUCA, the RNA world is a potential description of what you might have at, say, step 5 or 6 in the 12-step sequence. There are no viable explanations of how you get to that point, and no viable explanations that tell you how to get from the RNA world to our current system. You don't have a code, you have no proteins, you don't have DNA and you don't know how it would come about in the first place. Proponents of the RNA world, or other precursor systems, are prone to repeatedly using Genesis 1:3-like statements. Obviously, they don't say "and lo, the great Dawkins breathed on the dust and a protein appeared." They cloak their assertions with scientific jargon, sometimes pages long. If you Google "origin of life theories," any of the articles that come up will usually list about seven different theories. When you read each of them, you will see these kinds of statements "and then proteins took over from RNA." Without exception, these articles, along with books on the subject, and even papers in scientific journals, fail to adequately explain how the various chemical, statistical and conceptual hurdles associated with each step are overcome. They just say that somehow it happened. Genesis 1:3.

Scientists have been doing this since Darwin first published his theory. The first iteration was the theory of Ur-slime (originally called Urschleim and proposed by Ernst Haeckel). This was pushed by T.H. Huxley, Darwin's rabid bulldog. It hypothesized that life originated from a primordial pulsating mass of ooze, or slime, that possessed some unknown unique quality, producing the first living organism. It was presented using the same type of scientific jargon that we see today, but beneath this façade was nothing but gunk. But the jargon did the job and some people believed it because it sounded clever. The theories touted today are, in essence, no different but have just added new layers of sophistication to their jargon to better disguise the fact they are total garbage.

For example, in his book *Life Ascending* (2009), Nick Lane, a biochemist from University College London, and a science writer, says the following on the subject of the RNA world:

"And proteins, above else, are the masters of metabolism. It was inevitable that they would ultimately supplant RNA. But, of course, proteins didn't just come into existence at once; it's likely that minerals, nucleotides, RNAs, amino acids and molecular complexes (amino acids binding to RNA, for example) all contributed to a prototype metabolism. The point is that what began as simple affiliations between molecules became, in this world of naturally proliferating cells, selection for the ability to reproduce the contents of the whole cell." p54-55

Do you see an actual explanation of how proteins came into existence? This is a classic scientific Genesis 1:3-type assertion with some fairy dust of "molecular complexes" sprinkled over the top. Atheists like Nick Lane sincerely believe that life was the result of random processes, so fill their claims with assumptions of events based on this understanding of our origins. Websites, popular science writing and even scientific journals are riddled with this kind of wishful thinking. Most people gobble it up unwittingly because they do not know better, or do not want to know better. But please, even if you had very little understanding of DNA, protein and codes before reading this book, look at that passage and assess whether there is actually a sound theory, or, in light of your recently acquired knowledge, just Genesis 1:3 assertions.

There is no sound theory. The RNA world is a red herring[78], or a sleight of hand, or perhaps most accurately - a polished turd. On close scrutiny, it is not a plausible theory, no matter how much it is made to sound plausible. It is a distraction from the central questions of life's origins and is used to create an illusion that a really clever answer has been generated. But it hasn't, and this is the very best that the random processes argument has in its support.

4.6 SUMMARY OF CHICKEN AND EGG PROBLEM

So has what we have discussed helped or hindered the establishment's case?

- DNA is an arbitrary code and the codon assignments have no apparent physical or chemical connections to the amino acids they code for.
- The chicken and egg paradox is intact. DNA is utterly dependent on proteins for its code to be translated into amino acids, and proteins are utterly dependent on the DNA that codes for them in the first place. The two are completely interdependent, pointing to a *de novo* appearance, which would be impossible by random processes.

---

78   Wikipedia: A red herring is something that misleads or distracts from a relevant or important issue. It may be either a logical fallacy or a literary device that leads readers or audiences towards a false conclusion.

- There is no evidence of a precursor world in which either proteins existed on their own and were able to store information and be... proteins, or of a nucleoside world that did both. RNA in the ribosome is not a biochemical fossil pointing to an RNA world.

- There is no way of conceiving how this paradox could be overcome using random processes, and therefore the paradox is strong evidence against random causes.

- There is no explanation of how a code for an entirely different, unrelated, chemical system came to exist in DNA. There is no connection between the code and the output.

- Just as Dr. Wedgwood's machine could only be the result of intelligence, there is much about the DNA/protein code and translation system that strongly infers intelligent "interference." But I don't use that as direct evidence, so I won't count it here.

So here are the scores. Are you starting to get the feeling that the establishment is on dodgy ground?

	RANDOM PROCESSES		INTELLIGENT INITIATION	
EVIDENCE	FOR RANDOM PROCESSES	AGAINST RANDOM PROCESSES	FOR INTELLIGENT INITIATION	AGAINST INTELLIGENT INITIATION
AMOUNT (0-10)	0.5	6	0	0

5. THE FROZEN ACCIDENT: IT MIGHT BE FROZEN, BUT IS IT AN ACCIDENT?

5.1 INTRODUCTION

Francis Crick, one of the co-discoverers of the DNA structure, was troubled by the whole question of the origin of the code from early on. Can you remember the headline that I mentioned in an earlier chapter from NBC, in which ingredients of life may have been brought to Earth in space dust? Crick recognized soon after the discovery of the DNA code that the Earth hadn't existed long enough to allow life to come into being. He recognized that whether the Earth had only been able to support life for 6,000 years, 150 million years, or even a billion years, much longer time frames would have been needed for the DNA code and its accompanying translation machinery to have come into being by random processes.

To overcome this problem, he created a theory called panspermia. This required the traveling across space of pieces of DNA or RNA in comets or meteorites that crashed onto the surface of the Earth. This would have given the universe another 10 billion years in which it could have created life somewhere else (if you subscribe to the majority view that the universe is roughly 14 billion years old, and that life appeared on Earth around 500 million years after it first formed). But from the calculations in the numbers section, it is clear that even an extra few tens of billions of years would not be enough time to generate a single protein by random processes in this universe. Also, we learned that delivering biological material by smashing it into the Earth and creating an explosion that vaporizes rock is not really the ideal delivery mechanism; in reality, all we are doing with panspermia is just moving the problem somewhere else.

Crick also coined another term that highlights the challenges of proposing a theory for the "evolution" of this system. He called it the frozen accident.

The DNA code and associated translation machinery are universal to all life and have been since very soon after the earliest life forms came into existence. No other code or system has been identified in living organisms. This system is at the heart of life and precedes any subsequent evolution of species. Neo-Darwinian evolutionary theory proposes that mistakes made by this system cause the adaptations that enhance an organism's chances of survival. However, the one thing that appears to have never evolved is the system that lies at the heart of it all: the DNA code and its translation machinery. Therefore, the system appears to be "frozen" as it is.

Why is this?

Firstly, we need to understand how things evolve, at least from a theoretical Darwinian perspective. Some quick things to remember/note and key principles:

1. All the information for a living organism is contained in the genome of that organism. This is the entire sequence of all the nucleosides contained in our DNA. Every cell in our bodies contains this inside the nucleus.

2. Within the entire sequence of the genome are shorter sequences of nucleosides called genes. These code for the proteins that we discussed previously, and they make life happen.

3. In sexual reproduction, the genes from both parents are used to create the new being, giving a mixture of inherited characteristics. I won't go into this.

4. Cells, from human cells to bacteria, divide or make copies of themselves, sometimes many, many times. During this process, the DNA is copied. Occasionally (very rarely in humans), the copying machinery makes a mistake and the "daughter" cell's DNA will contain what is called a mutation.

5. Mutations are random mistakes. Most are either harmful, resulting in the cell dying, or perform badly, or have no effect. Very occasionally, a mutation may confer benefit, which gives the cell or being a survival advantage.

6. The system has no foresight. Mutations are random and survival outcomes are purely a result of good or bad luck. These natural processes do not have a goal or intention. They do not have a mind.

To understand how this kind of evolution works, look at the following two nucleoside sequences that could form part of a gene:

TTAGTAATCACGGGT is the original sequence

And

TTAGTAATCTCGGGT is the "mutated" sequence, which was incorrectly copied from the first sequence.

There is one small change: an A, adenosine, has been substituted with a T, thymidine. While the translation machinery in advanced organisms is excellent at spotting errors, occasionally one might slip through, given the trillions of pieces of code that are copied every day.

If you read the sequence as triplets, and translate the sequence using the universal DNA code, the original DNA sequence translates into the amino acid sequence:

leucine – valine – isoleucine – threonine – glycine

Whereas the second sequence translates to:

leucine – valine – isoleucine - *serine* –glycine

If you remember from previous sections, different amino acids have different properties, and these differences cause the protein to fold into certain shapes and perform very specific tasks. Changing a threonine to a serine could significantly alter the properties of the protein, especially if the change, or mutation, occurs near the "active site" of the protein. (The active site is where some proteins perform work on other chemicals.) It is important that these sites have a specific shape for the specific chemicals they are supposed to work on. Below is a very crude representation of what might happen with the change in sequence above. (Remember, the blobs are just for illustrative purposes. Proteins are linear chains that fold into specific structures, and blobs are just one way of representing them).

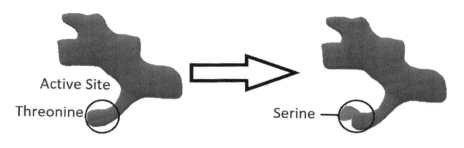

In this example, the change alters the shape of the active site. This would most likely make the protein less efficient at performing its task, but occasionally such a change might actually enhance the ability of the protein to perform the task. If a specific chemical is supposed to sit in the active site, the shape of the chemical could be more perfectly suited to the new shape and the protein is able to function more efficiently. This will confer a "survival advantage" to the "mutated" protein, and therefore the new genome will be better and more likely to survive.

At least that is what Neo-Darwinism would have us believe. ID proponents, and an increasing number of mainstream scientists, are questioning whether this kind of mechanism could account for the large variation of species and phenotypes observed in nature. Epigenetics, an exciting area of cutting-edge science, is observing "real-time" evolution that challenges Neo-Darwinism. But this lies beyond the scope of this text, as they all refer to events that happened after the appearance of DNA and its translation machinery.

One indisputable fact is that these types of mutations do cause evolution to occur on a micro scale, and in microorganisms such as bacteria or viruses, they can cause resistance to drugs. They also cause cancer when mutations cause cells to divide more than they are supposed to.

As mentioned previously, I have spent many years working in HIV, and these types of mutations presented a huge barrier to the generation of successful therapy. Understanding how this happened will help us understand the reason why the DNA code is frozen.

5.1.1 KNOWLEDGE UPGRADE

Modern antiviral drugs are often designed to compete with natural molecules at viral protein active sites. Viral proteins are proteins specific to the virus and help construct new viruses. By mimicking the natural molecules that the viral protein is supposed to be working on--such as the nucleosides A, C, T or G--the drugs (e.g. AZT, a manmade nucleoside) fool this viral protein into accepting it instead of the natural one. Using the man-made drug will disrupt natural viral reproduction and eventually kill the virus. We can use the blob to illustrate how this might work.

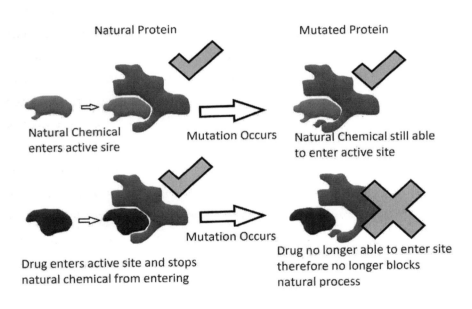

Natural Protein   Mutated Protein

Natural Chemical   Mutation Occurs   Natural Chemical still able
enters active sire   to enter active site

Drug enters active site and stops   Mutation Occurs   Drug no longer able to enter site
natural chemical from entering   therefore no longer blocks
                                                            natural process

Imagine that the natural viral protein (blob) on the left accepted both the natural and manmade chemicals into its active site. The "natural" chemical is the one that the protein is supposed to be using to build a new viral particle. The drug however, which fools the viral protein into thinking it is a natural chemical, would enter the active site and perform whatever chemistry it was designed to so that the process of building a new viral particle is disrupted. For example, it could permanently block the active site of the viral protein, taking that protein out of service, or it could be incorporated into a new viral particle that the protein was building and make it faulty.

Now viruses do not have the same level of accuracy as DNA when it comes to making copies of themselves. In fact, HIV produces millions of mutant viruses in an untreated, infected individual every day. It is sloppy, and in this instance the virus spits out a mutation that results in the mutant protein on the right.

This new mutant protein is still able to process the natural chemical in the way it was designed to, and therefore produce fresh copies of the virus. However, the change in the shape of the active site from the mutation means the manmade drug no longer fits in the active site. The drug is no longer able to block the protein and kill the virus. This generates a clear survival advantage for the viral particle that has the mutant genome when the drug is present. Eventually, this new, mutated virus will grow out and predominate. This virus is now "resistant" to the drug, as the drug no longer works. The virus has "evolved" to develop drug resistance.

This is exactly what happened in the early days of HIV with AZT, DDI and other nucleoside antivirals (or antiretroviral, as they are properly called). People would get better for a short while, but then the virus would mutate. The mutant virus would become resistant to the drug, proliferate, destroy the immune system and the patient would die of an opportunistic infection.

So why is it that people infected with HIV today very rarely develop resistance? In fact, why is it that someone diagnosed with HIV in the early stages of the virus is likely to live at least as long as someone who isn't infected?

I will answer that question in a short while, as it is a perfect illustration of the problem central to the frozen accident.

Hopefully this has given you an insight into how evolution, or at least micro-evolution, happens. In summary:

Random DNA mutation in genome > change in protein > survival advantage > new genome becomes dominant

So evolution is a random, natural process. The mutations that occur in DNA are random mistakes. These mistakes are translated and change the resulting proteins in the daughter cells or organisms. If these changes provide sufficient advantage, these new cells or beings, which contain the new genome, predominate.

There is the only one form of feedback. Survival. The fittest genome survives when there is competition for resources. The old genome, if it is less fit, is "retired."

All of the effects of mutations are downstream. In other words, the next generation of cells is the one that is affected and either survives and thrives or dies off because it is faulty. There is no upstream feedback. There is no intention. There is no conversation between molecules, only random mutation and survival.

5.2 SOLVING THE FROZEN ACCIDENT BY WORKING BACKWARDS

So far, most of the theories that have been discussed, such as those around generating starting materials, or complex biomolecules, are a bottom-up approach. How could things start? In the case of the RNA world, it was a start-in-the-middle approach. As I have shown, and is the case, no one has been able to answer the central questions using either approach.

But what if we try and work backwards? This is often what detectives will do when trying to solve a case. They will try to find out the last movements and then trace the case backwards, filling in gaps.

It is when we try to do this with the question of the origin of DNA that the issues surrounding the "frozen accident" problem come sharply into focus.

What could have been the previous step, the state of DNA and the system that existed immediately prior to the one we have today (and have had since very close to…if not, the beginning)?

Reminder of the system:

Universal code in which triplet combinations of four nucleosides are translated by twenty different ARS proteins (with help from RNA) into twenty amino acids.

5.3 A SMALL STEP BACKWARDS

What would the predecessor of this system be if we took a small step back? The simplest change would be to have a system which only coded for nineteen amino acids. To go from this to the current system, you are only adding one more amino acid to the dictionary. Sounds simple enough, but is it?

Let's just think about the ARS protein. Remember, this is the protein that lies at the heart of the translation process. It is hundreds of amino acids long and has a high degree of specificity for the molecules that it works with at the active sites where the work is done. This specificity is achieved through a precise and unique sequence of amino acids that is able to accurately identify these molecules. If the sites are not precisely specific for these molecules then the ARS will not accurately translate the code, and the organism will die. There are two of these highly accurate sites, one for the anticodon of the tRNA molecule and one for the amino acid that the ARS will eventually join to the tRNA.

Let's say you are able to recycle one of the existing nineteen ARS that currently works on a similar amino acid. What I mean by that is if there was an ARS that worked for one amino acid, it is possible that, with a few alterations at the amino acid active site, it may work for a very similar amino acid. However, despite the similarity, you would still need to change the active sites on the ARS to accommodate the new shape. These changes in shape would require substitutions/additions/deletions in the existing chain of amino acids that make up the ARS. The same applies for the tRNA active site of the ARS.

One change in these active sites would probably not be enough. Let's be really generous and say that you would only need to make two changes in each active site--two in the active site that binds to the tRNA site, and two in the site that binds to the amino acid to make your twentieth ARS, making a total of four changes. What would be involved to achieve this?

Firstly, this would have to happen in one step. Intermediate steps would create no advantage, as they would either only bind the new tRNA, or the new amino acid, not both. So you would need to have all four changes in the amino acids, at the correct positions, at the same time. This would require corresponding changes in the nucleoside sequences in the original DNA, in the correct places at the same time, in one solitary cell. How likely is this?

Let's go back to HIV to get a tiny glimpse of why this is so hard.

Someone newly diagnosed with HIV today will live at least as long as someone without HIV. The reason for this is that we have developed *combinations* of drugs that permanently keep the virus under control, and crucially stop resistance developing. How, and why is this relevant?

Prior to receiving drugs, the body of someone infected with HIV will contain billions of HIV virions (viral particles -- think of cells) using host cells to replicate like crazy. This is represented in the diagram below by the first bucket. As I said, the virus is sloppy, and constantly produces mutations. These mutations mostly make the virus less fit, and they do not grow out, and the *wild type* virus predominates. However, these mutant virions will persist in tiny amounts in the background.

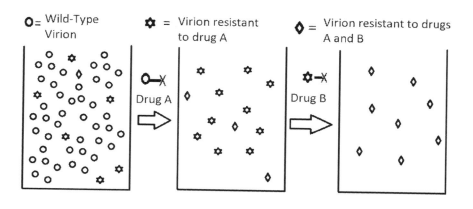

When you add a single drug, e.g. drug A in the diagram above, to this mixture of virions, it destroys the wild type strain within a matter of weeks. However, there will be a number of virions amongst the billions present in the patient that will be resistant to drug A (remember the blobs). In these resistant virions, the target protein is still able to process the natural chemical, but not the drug. With the competition from the wild type strain removed, the mutant virions are able to successfully replicate, and eventually grow out and become just as prolific as the original virus, creating billions of new virions with lots of new mutations.

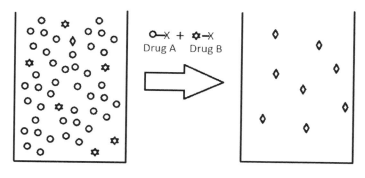

What if we now add a second drug, drug B, that targets the virus in a different way, but still we keep drug A in the mix? The process is repeated. The predominant virus now has two mutations.

You can keep doing this, and we did, with the virus picking up mutation after mutation until we ran out of new drugs and the patients died. Researchers then tried adding two drug combinations together at the start of therapy. This seemed to work more effectively at first, but eventually, albeit after months rather than weeks, a new resistant strain would emerge. This invites a couple of questions. Why didn't two drugs work and why did it take longer for resistance to develop?

Two drugs didn't work because there were still virions present at the start of therapy that had resistance to both drug A and drug B. The reason that it took longer for the resistance to emerge was that there were far fewer virions present with resistance to both drugs. While there may have been hundreds of virions that had resistance to drug A only, there may have been just one or two that had resistance to both drug A and drug B out the billions of virions present. Given this, it would have taken many more replication cycles to reach the kind of numbers to be detected by tests. What is important to remember is that the mutations must be present in the same virion for the virus to become resistant. In other words, one genome from one virion must contain both the mutations for this to happen and the virion must remain competent. The chances of this happening are much smaller than just having one successful mutation present, so this is, in fact, all about numbers, or statistics.

In 1997, triple combination therapy was first used. This changed things forever. It turned out that, by using three drugs, the virus could be permanently kept under control. Resistance did not occur (provided the drugs were taken properly). Why is that? I suspect you can guess:

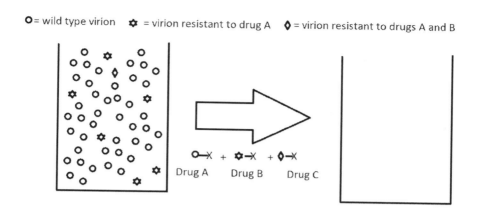

While it was statistically possible for there to be a small number of viable virions present that had two drug mutations, it was exceptionally unlikely for there to be a single virion with three successful drug mutations present. Remember, mutations are mistakes, and most mistakes are not helpful or fatal, so the chances of getting one that is helpful is relatively small. The odds of a virion containing three successful mutations appearing before exposure to any drugs, even accounting for the billions of cells in each infected individual, and the millions who have been treated is incredibly low. This is why triple combinations work. There are possibly only a handful of patients, if that, out of the tens of millions who have been infected by Wild type[79] virus who had a triple-drug-resistant mutant virion produced via random means

---

79   Wild type virus is the most common and is the one that evolved without selective pressure from anti-retrovirals (ARV). A significant number of newly diagnosed HIV patients are infected with a resistant strain from partners who have developed resistance to one or more ARVs. This immediately reduces the options available for this patient. Outside of these cases, it is technically possible that patients who had wild type virus when they are first infected, could produce a virion that has triple drug resistance, i.e. 3 viable mutations that are resistant to 3 different ARVs, without being exposed to any of these ARVs through random mutations. However, given there are no documented cases of this, it is extremely unlikely, showing the difficulty of producing 3 successful mutations in a genome in one step.

in their viral reservoir before starting treatment for the first time.

So what does this have to do with the problem of generating a new ARS protein?

Being generous, I guessed that you would only need four successful mutations to generate a new ARS capable of translating a new amino acid to expand the code from nineteen to twenty amino acids. Having seen the HIV example, I hope you can see that the chances of this type of change occurring in one replication cycle, even given the trillions of organisms that have existed, is extremely small[80]. However, simpler changes in the code have indeed happened, as there are thirty-four known codon reassignments, some of them present inside you in mitochondria. Codon reassignments (usually reassigning a stop codon to an amino acid or vice versa) require the least amount of changes (mutations) to the code in one step; you are not expanding the code. It does show that the code has a bit of flexibility, but the wiggle room for changes is very small. I discuss this in much greater detail in the appendix.

Expanding the code by just one amino acid is a whole different ball game. Not only would you need a new ARS protein, you would need to change the DNA polymerase in the nucleus to be able to read the new code. You would need new tRNAs. You would also have to ensure that the new amino acids that are added enhance the new proteins that are made, rather than damage them; the positioning would be vital. In other words, you would need dozens of random, simultaneous mutations all in ONE new copy of a genome, all of which would have to be beneficial, just to expand the code by one amino acid. The odds against this are extremely small.

But you still haven't overcome what is the biggest problem by far. Where does the new amino acid come from, and WHY does the new code for it appear? This is no longer a statistical hurdle, but a conceptual hurdle, and one

---

80 Some argue that the code has expanded citing the two non-standard amino acids found in nature. This is discussed in the following pages and in greater detail in the appendix, along with the reasons why, in my opinion, it does not support the belief that the code evolved.

156

that appears very difficult to solve. Think about it for a second. The system is biomechanical in nature; it has no cognitive abilities. It only changes by making mistakes. The system is not *aware* of the existence of a new amino acid, even if the amino acid exists.

WHY on Earth would it generate a code for this new amino acid?

Suggesting that we didn't know about processes and molecular capabilities billions of years ago does not wash. We know exactly what we have now, and we know exactly what would be required to go back one teeny tiny step, and therefore we understand exactly what would be needed to go from that previous system to the current one. We know that biochemical systems do not have foresight. We know that they would not "know" that adding a new amino acid would help. This is compounded further by the question of the generation of an entirely new codon to code for this new amino acid and then knowing where to put this new amino acid in the proteins it creates without completely messing up those proteins.

When you start to think of the system in these terms, you begin to understand the extreme difficulties facing the expansion of the code by random processes. However, and it is a point of debate, it may have happened.

There are two amino acids used in nature in a few organisms in addition to the standard twenty used by the vast majority: selenocysteine (Sec) and pyrrolysine (Pyl). The first, Sec, is not a genuine addition to the code as it is not actually coded for in the DNA. However, the means by which it is included is fascinating, and does require changes in the DNA. The best way to describe it is a post hoc modification. Ply, used exclusively by organisms that live by volcanic vents and use methane in their metabolism, is a genuine addition to the code because it has its own ARS and tRNA. I discuss the details of these two in the appendix but will briefly summarize the conclusions that can be drawn here.

- Firstly, there is only one genuine case of expansion in the code, if indeed it was expansion. This actually exemplifies the points I have been making. Out of all of the trillions and trillions and trillions of bacteria and archaea that have existed, we only know of one in which the code may have expanded by one amino acid. Such are the statistical barriers facing an expansion, that we do not know of another example out of all of the many species that we have investigated. Of course, there could be others that we haven't found.

- Secondly, and this is where the debate focuses, is Pyl evidence of an expansion or maybe a contraction of the code? Both Sec and Pyl exist in archaea and bacteria, meaning they were either present in LUCA or appeared very soon afterwards. This is debated in the appendix. Suffice to say, in all the years since life split, there has been no expansion of the code that we are aware of.

Aside from the fact that expansion would need to have happened many times to produce the current system, there is another issue with the small-steps-backwards approach: how many small backward steps could you take before you would no longer have a viable system? There is a minimum number of amino acids that would be required for a functional protein, and it is not one or two; it is in double figures. So even if the system could evolve one amino acid a time, you reach a cliff edge beyond which no more single steps backwards would be viable. At that point you would face the conundrum of how to get a 10-14 amino acid system *de novo*. But this is a moot point, because there is an even more precarious cliff edge.

5.4 A BIG STEP BACKWARDS

So far, we have looked at the code being frozen as a code for twenty amino acids, and how difficult it would be to take a step back to a nineteen amino acid system and add just one new amino acid to the code. What if we look at a simpler coding system? While there is debate about whether the code evolved from twenty to twenty-one amino acids in an ancient, pre-LUCA organism (or LUCA itself), no reasonable scientist in his right mind would believe for a second that a triplet coding system would appear *de novo* by random processes. This would require the simultaneous formation of the code for hundreds of components by random processes in one event. It is beyond the realms of even the most ardent atheistic fantasies, which is why it is never postulated in scientific literature, not even high school textbooks. It would have needed to have come from a simpler system. A doublet coding system is the obvious choice. You would have fewer amino acids to code for (4X4 combinations), but you could still just about produce some viable proteins with sixteen amino acids, so theoretically it is a system that could have existed.

But how would you go from a doublet system to a triplet system in one step?

Answer: with multiversal difficulty. You would have to change every single ARS, along with reassigning the entire code at the same time and then rewrite the code for every single protein in triplet code that had previously been coded in doublet. At this point, it is objective to say that this is just beyond ludicrous from a statistical perspective. Could it have happened in a stepwise manner?

No. It couldn't happen in a stepwise manner. You can't have half the genome intended to be read as a doublet code and half as a triplet code. How would the translation machinery know which is which?[81]

To understand why this is the case, let's go back to Dr. Wedgwood's code. Dr. Wedgwood, having lined his pockets in Silicon Valley, has decided that producing code for sock puppet farms was not his bag after all. Not realizing that it wasn't only Russians but also the CIA who used these farms regularly, he decided to head back to the "noble" work of the government. He is immediately put back in charge of the project he created, only he decides that a triplet system is no longer adequate. After all, he needs to justify the huge pay increase he received compared to his former tenure at the agency. He decides to expand the code to a quartet code, i.e. sequences of four letters now code for one Arabic symbol. He creates a new dictionary, hands it to the operator in Langley, and walks out the door before the operator is able to ask a key question:

"Did you change the machine as well?"

The operator scratches his head. He has an important message that needs to be communicated to the Iranian asset immediately, something along the lines of, "You've been rumbled mate; they're on their way to kill you."

---

81   One of my editors, who was extremely helpful in that not only did she help me enormously with structure and tone etc. but also because she was clearly at least agnostic and possibly verging on atheist, peppered the manuscript with the kind of questions I would expect from people holding such beliefs. It was at this point that she posed the ultimate atheist cop-out which nearly induced an Alex-Jones-style fit. I hope she doesn't mind, but I will quote: "no offense, but is it truly impossible or is it possible science/math simply can't explain it yet?" This is the science of the gaps argument; science is so clever that it will answer the question one day. Firstly, we are working with the data that we currently have, unknown future scientific discoveries don't yet exist and so can't be evaluated. Secondly, we have to go with science's track record on this subject so far. Over fifty years of some of the brightest minds on the planet contemplating this subject, and nada. While science has a good track record on many things, when it comes to origins of life, it is rubbish and even the scientists themselves, like Sutherland, admit this. Thirdly, we know a lot about this system--about chemistry and biochemistry. We know what is possible and what is statistically impossible, what is plausible and what isn't, and in this instance, I added the CIA analogy to show why it is impossible.

What should he do? Being a government employee with aspirations for leadership, he decides to fudge things. He decides the best approach would be to send half the message to the operative in Iran using the old triplet code and the other half using the new quartet code. Can you see the problem?

Going by Dr. Wedgwood's track record, it is likely that the machine wasn't upgraded and still only reads the code in triplets. It can read the input which is a continuous stream, just fine: ACCTGGCAATTT, but it would then always break it down to ACC TGG CAA TTT, whereas the sections that were inputted using the quartet code should have been read ACCT GGCA ATTT. The two have completely different meanings. Likewise, if Dr. Wedgwood had got up in the morning and, for once, worked tirelessly from dawn to dusk, and actually changed the machine at the same time as the code, then the parts where the operative had used the old code would be read in quartets instead of triplets. Given that the letters are in a continuous stream, there is no way of differentiating which part should be read in a different way. You either have one or the other, and a machine that reads only one system.

The message the Iranian asset received read something like this: "Kill cow on monocycle mate." The next day the asset was found dead on the floor with a scrunched-up piece of paper next to her.

To apply this to our translation system, you couldn't expand the code from a doublet to triplet system in steps, since the doublet ribosome and tRNAs would be incapable of distinguishing which parts were supposed to be doublet or triplet. Therefore, the only way you could "upgrade" to a triplet system from a doublet system would be a wholesale replacement of the entire system and a complete rewriting of the code, not just for the translation system, but also for everything else in the organism, as the old code was written in doublets. Yes, in one step, that is impossible by random processes, and as likely to happen as twenty-two cows jumping onto the moon, playing a soccer match, with halftime entertainment from the Elvis impersonating cow, and having tired legs, building a space shuttle to return to Earth instead of jumping.

Science writers who are also respected scientists, like Nick Lane in his award winning and critically acclaimed[82] book *Life Ascending*, use a scientific Genesis 1:3 and pile a heap of pixie dust on top, hoping no one will notice that a quasi-religious proclamation has been made. Using endless conflations of bad science (suggesting that codons bind to their amino acids, which they don't) and incomplete theories, he creates an impression over three pages (49-51) that the problem is easily solved. However, like all of these explanations (of which there are few, because it is actually inexplicable) not one measures up to the required criteria of being chemically, statistically and conceptually sound. None of the issues raised here are addressed and no logical pathway proposed. The conclusion of this section of his book, the one that describes how you go from a doublet to triplet system, says this:

"This whole scenario is **speculative**, to be sure, and as yet there is **little evidence to back it up**."

Fine so far, at least that is honest, but then the next sentence reads as follows.

"Its great virtue is that it sheds light on the origin of the code, taking us from simple chemical affiliations to a triplet codon in a **plausible and testable way**."

Compare the bolded underlined sections. Talk about internal contradictions in juxtaposition! The two sentences actually appear in that sequence. It's like saying "Everything I just told you is made up nonsense. It is a good theory because it shows us how it could happen, and is factual."

Whether it be the fake news establishment media, popular science writers or even scientific journals, all theories that try to explain the appearance of the code by random processes are lacking in evidence or even conceptual integrity. They are hollow brain farts. There is no exception to this rule.

---

82   The Royal Society gave it an award and the *Guardian* and *Observer* newspapers in the U.K. gave it rave reviews.

5.5 SUMMARY OF THE FROZEN ACCIDENT

Many of the greatest minds in this field, like Francis Crick, have observed that the DNA coding system is a frozen accident. This is what the evidence tells us. We don't know of any other biological coding systems. The DNA triplet code is universal, and always has been. No one has been able to generate a viable theory which shows how you go from a simpler system to the current one.

More importantly, no one has been able to figure out WHY a code would expand.

The idea that a triplet system appeared *de novo* is preposterous to a 'rational' scientist, and yet there is no conceivable way by which we could go in a stepwise manner from a simpler system.

Scientists shrug and say it is unknowable or unsolvable. Those are the wrong terms. The correct term is the one that I used in the previous sentence. INCONCEIVABLE. This is factually correct. No one can even conceive of how random natural processes could generate this system. We know where we are very well, and yet we can't even figure out how we got here, even from something similar. Could that be because random processes are not the answer? Could it be the code is not an accident?

Time to update the scores:

- Evidence that random processes generated the code. None
- Evidence that random processes could not generate a code. Lots…it is inconceivable. Add two points to the 'against' score.
- Evidence for intelligent input…getting very warm now…but not yet.
- Evidence against intelligent input. None.

EVIDENCE	RANDOM PROCESSES		INTELLIGENT INITIATION	
	FOR RANDOM PROCESSES	AGAINST RANDOM PROCESSES	FOR INTELLIGENT INITIATION	AGAINST INTELLIGENT INITIATION
AMOUNT (0-10)	0.5	8	0	0

6. NOT HOW BUT WHY?

Before I come onto the final section, the evidence for intelligence, I want to just go into the "WHY?" question one last time. This specifically relates to the subject of information.

This is a quote from the BioLogos blog[83]:

"Thirdly the amount of information needed to describe the environment of the pre-biotic Earth as a whole was much greater than the amount of information needed to describe the first living organism, Fourth, on the early Earth there was a steady stream of orderly energy (sunlight) and a

---

83   BioLogos appear to be of the view that God set in motion all the laws of the universe, and that it was his intention that these laws would result in the formation of life. It is interesting because physicists are also moving to a similar conclusion, albeit without a God, something I will come to in this section. Anyway, BioLogos argue on their website that life was not a specific act of intelligent intervention outside of natural laws, but was an event resulting from natural laws that had been intelligently invented and that would one day lead to life. For those outside of the world of Christian apologetics, you might wonder what's the difference. To those of us in that world, the answer is: everything.

constant thermal jostling and mixing of the molecules. These are exactly the sorts of conditions which enable the creation of information needed to self-organize real, complex objects."

This is one of the many fallacies I have found on the subject of how or why information would get into a natural system by random processes. Like so many, they are muddling complexity with information. Below are two relevant definitions for "information" from the Meriam-Webster online dictionary:

1. the communication or reception of knowledge or intelligence

2. a (1) : knowledge obtained from investigation, study, or instruction (2) : intelligence, news (3) : facts, data b: the attribute inherent in and communicated by one of two or more alternative sequences or arrangements of something (such as nucleotides in DNA or binary digits in a computer program) that produce specific effects c (1) : a signal or character (as in a communication system or computer) representing data (2) : something (such as a message, experimental data, or a picture) which justifies change in a construct (such as a plan or theory) that represents physical or mental experience or another construct

Meriam-Webster has obviously been very helpful by actually specifying the kind of information that is in DNA. Information, in this instance, relates to an attribute defined by the sequence in, or arrangement of, the DNA that can be communicated. The dictionary defines it alongside computer code. The DNA is not the information, just like computer chips are not the information. The complexity of a DNA molecule is not the information; the information is the attribute or something that is described by the sequence of the nucleotides.

Complex molecules or structures that are able to form from purely natural processes may have the capacity to contain information, however, without intelligence inputting the information, that is as far as it goes. Their complexity alone is not information.

To understand BioLogos's mistake, you only have to look at the first sentence which contains the statement "needed to describe." The complexity of the pre-biotic Earth only becomes information in the context we understand the word, or that is described in Meriam-Webster, once the structures are described, or observed. A lump of coal or a diamond is not information, but the description, or observation of it is. Visual observation is, itself, information since the image that forms on your retina is converted into electrical signals that are transmitted along the optic nerve to your brain. The brain decodes the electrical signal and you "see" an image; this is information.

An even more hideous statement follows immediately after the first sentence--namely, that adding energy to a mixture of molecules causes lots of reactions, which might lead to self-organization and the creation of information.

This is complete chemical nonsense with regard to self-organization, as I showed in earlier chapters. Moreover, any computer scientist will tell you that randomly scrambling components of a code is more likely to lead to less information than more information. You may indeed get greater complexity, and it might require a lot of information to describe the mess, but the complexity itself does not hold information of the nature that is present in DNA.

This is the critical distinction that needs to be understood. The kind of information we are talking about conveys a specific message with a specific meaning.

Let's use DNA as an example. Consider the following sequence of nine nucleosides:

A-A-T-G-A-C-C-C-G

Now imagine there are two scenarios. The first scenario is one in which this sequence appeared in a world where there was no DNA code. There was no translation machinery, the sequence was purely random. You could describe this as follows, "a sequence of nine nucleosides starting with A followed by..." The description is information, but the sequence itself does not contain information that actually *informs*.

Now consider the same sequence in the world in which we live. The physical description would contain the same amount of information, and the level of physical complexity is identical, however there is an additional abstract quality or characteristic that this sequence possesses that the first one doesn't. Not only would it require more information to describe it, but it actually conveys information itself. This sequence is actually *describing* something completely unrelated to its physical or chemical characteristics. It is describing the sequence of amino acids: Asn-His-Pro (asparagine, histidine and proline). This is abstract information derived from an arbitrary code.

The fact that scientists at BioLogos and others can't seem to be able to differentiate this quality from physical complexity means they are either not as smart as they claim, or too smart to be able to see the obvious (fleas on an elephant), or they are being deliberately disingenuous.

It seems that some serious scientists get stuck on one definition of what information is, and completely misunderstand the nature of the kind of information that is contained in a code or language.

Without getting into the details of information theory, it is accurate to say that DNA possesses a higher level of information. It is information that has abstract meaning...meaning that is entirely unrelated to the qualities of the medium of the information and meaning that is only understood through translation. Namely, DNA, the nucleoside language, contains specific information for proteins, the amino acid language.

Now let's move to the WHY issue.

Why on Earth would DNA or RNA develop a code for a completely unrelated chemical system?

Now the answer given by atheist scientists is that proteins are much more efficient at performing processes than nucleoside biological "machines," such as those found in the RNA world and, at some point, the nucleoside system developed the code for amino acids.

That answer to the why question is a logical, intelligent answer, and intelligence would indeed be required to initiate this response to the problem of efficiency, but random natural systems do not have intelligence. They have no foresight, no goal, and no knowledge. They would not know that proteins are better, and they would not have a "desire to be better." So WHY would they do it?

Moreover, why would a nucleoside system code for a completely different system at all? The conceptual hurdles are insurmountably vast for a system with no intelligence, because there are two levels of abstraction in the DNA code. What I mean by that, is the final meaning of the code is unrelated in two ways from the original code.

How so?

The first level is that three nucleosides code for one amino acid. There is no physical or chemical relationship between them, no reason for the sequence to code for that amino acid[84]. They are not logically connected in a physical way. That is one level of abstraction. It is precisely the same as a banana coding for the word elephant.

The second level of abstraction is the final resulting function of the protein in relation to the DNA code. The physical shape and electrochemical properties of the protein that perform the function have no relationship to the code. What do I mean?

Let's imagine an RNA molecule that performs a specific function. It is acting like a protein, as RNA molecules are able to. The sequence in that RNA gives it a shape etc. that allows it to perform that function. The RNA, being a sequence of nucleosides, by default also codes for amino acids. Should the resulting

---

84   In the appendix I cover aptamers, rRNA, and attempts to link frequency of certain nucleosides in key sites to codons. I discuss in detail the work by Yarus that looks into this.

sequence result in a functional protein, it would only be by an incredible fluke that the protein would have the same function as the RNA. There is no logical connection between the physical properties of the DNA or RNA molecule and the final function of the protein it codes for. The structural properties defined by the RNA code for the protein are completely unrelated to the structural qualities of the RNA.

The DNA code has two meanings. On one level, it is a code for amino acids, but on another it is an instruction for a function. The two are conceptually unrelated to each other, and neither is related to the fundamental physical properties or abilities of the DNA or RNA molecule.

The code is arbitrary and abstract, and the question of how this could appear is not answerable invoking random processes.

It is utterly inconceivable, which is very strong evidence that it couldn't have happened. In my view, this adds another 1 to the column of evidence against the belief that the DNA code could have come into being by natural random processes.

	RANDOM PROCESSES		INTELLIGENT INITIATION	
EVIDENCE	FOR RANDOM PROCESSES	AGAINST RANDOM PROCESSES	FOR INTELLIGENT INITIATION	AGAINST INTELLIGENT INITIATION
AMOUNT (0-10)	0.5	9	0	0

This concludes the assessment of all the evidence relating to the belief that random causes could have resulted in the appearance of DNA or life. Do you feel that the Canadian Governor General was right to sneer condescendingly at those who don't share her belief in random causes? Do you think the Canadian Prime Minister was being a tad unwise when he stated that she was a defender of the truth? Do you think the establishment has been misleading us on this important subject?

Now you might be thinking that I have loaded the dice; that I am biased or ill-informed. Given that I do not have the establishment seal of approval tattooed across my forehead, I can understand your concern about how these conclusions might have been reached. As I mentioned in my opening paragraph, changing a worldview is very hard, no matter how objective you might believe you are. But before you shake your head and tell yourself this is all nonsense, I want you to do two things.

- Firstly, as I suggested on a number of occasions, check whether what I have said rings true. If you Google "origins of life" and find coherent theories from A-LUCA, then I am wrong. If there are viable (chemically, statistically and conceptually sound) routes to proteins, or RNA, or DNA, or cell walls, or code, then I am wrong, and I apologize for wasting your time. But if you can't find anything, then maybe consider that there might be more than an ounce of truth between the covers of this book.

- Secondly, ask yourself whether that table of evidence regarding the random processes claim--namely, 0.5 in the evidence for and 9 in the evidence against, rings true with the graph that I recreated from the Sutherland paper? Does it ring true with the following statements from another 2017 review[85] that I discuss in the appendix:

"The origins of life stands among the great open scientific questions of our time. While a number of proposals exist for possible starting points in the pathway from non-living to living matter, these have so far not achieved states of complexity that are anywhere near that of even the simplest living systems."

---

85   Origins of life: a problem for physics, a key issues review; Sara Imari Walker 2017 Rep. Prog. Phys. 80 092601

And,

"The idea of information is itself abstract, but it must be the case that each bit of information is instantiated in physical degrees of freedom. "Information is physical!" in the words of Rolf Landauer. Whether fundamental or an epiphenomenon, the causal role of information in biology represents one of the hardest explanatory problems for solving the origins of life."

This latter paper was written by Sara Imari Walker, a colleague of Paul Davies, who produced an excellent book on the origin of life, and who was the first to float out a theory that life, and information in living matter, was the result of an as-yet undefined fundamental law of physics that drives these things into existence. That was proposed back in about 2000. This paper was written in 2017. Walker, and others, have got no closer to identifying such a law and yet she still claims physics will provide the answers. It is pure hubris on the part of physicists who recognize that no one else on the planet has a clue what they are talking about, so they can say anything they like and we will believe them. It is not unlike a father walking into a room and finding his wife and two teenage children puzzling over a pile of screws and wood with some instructions relating to the assembly of a set of shelves. The man grabs the instructions from his teenage daughter and says, "Don't worry, I'm here now, I'll have this done before you know it."

"But Dad…" the daughter starts.

"Uh uh!" the father says holding up a hand to stop further objections. "I have put lots of these together over the years. I know what I'm doing."

His wife rolls her eyes, gets to her feet and gestures for their kids to follow. They walk out of the room shaking their heads.

Two hours later, the air in the room has turned blue from the choice of words their father has used to describe his progress. He has his sleeves rolled up and is looking angrily at a wonky construction that bears no resemblance to the picture on the box. He starts unscrewing it to try another time; he will not be beaten.

Meanwhile, the mother and her children are sitting on their deck enjoying the late afternoon sun. She is sipping a glass of wine and the kids are playing on their phones. The daughter looks up from Instagram briefly.

"I feel guilty," she says. "We should tell him there were two pieces missing when we checked the contents."

The mother contemplates the suggestion. "Let me finish my wine first."

Having said this, the physicists may be on to something. There may indeed be a force that puts information into systems and gives them life. There may be laws that are designed to be manipulated at a quantum level by this force to generate physical, observable outcomes. But the nature of this force is not random.

Time to take the red pill[86].

---

86   One of my editors suggested that using this expression might be offensive to those who associate it with a pernicious group of US misogynists who denigrate women. Firstly, it is unlikely that the kind of people who get offended by an unintended reference to a small group of men would have the intelligence to get this far in the first place, so I doubt anyone will have been offended. Secondly, the Matrix is my favorite movie, and I will not stop using the term "taking the red pill", which is, and always will be, associated with freeing your mind from the lies of the world around us, just because a few "triggered" people might be offended. Thirdly, like millions of others, I am sick of overzealous, self-appointed members of the PC police destroying our ability to talk sensibly, and disagree respectively, with each other. In other words, get over it.

7. EVIDENCE OF INTELLIGENCE

Society has done a complete 180. It used to be that almost everyone believed that life was created by an intelligent being, or God. Taking the red pill meant meeting secretly with fellow enlightenment travelers and discussing the forbidden fruits of scientific knowledge that challenged the dominant cultural religious orthodoxy. Now, especially outside of the U.S., the opposite is true. It is counter-cultural to believe that a specific creative intelligence "breathed" life into lifeless chemicals. It is against the establishment orthodoxy to state this belief publicly, especially if you work in science-related fields or academia. Most people under the age of forty will have had barely any exposure to education that tells them anything other than the materialist story. For them especially, opening their minds to a different narrative may be a challenge. To do this, you need to have rebellious genes combined with a thirst for truth. Taking the red pill is not for those who wish to have it easy.

So far, you've not had to choose blue over red, or vice versa. All of the evidence we have reviewed relates solely to whether or not random processes could be responsible for generating life. The evidence is strongly against that, which raises uncomfortable questions, but you can still sit on the fence; you can still avoid the pill choice. This section is where all that changes. You will be forced to make a choice about the sum of the data. You will need to accept or refute the evidence for intelligent initiation of life. So what evidence is there?

I'm going to start with an analogy.

An explorer is exploring a region of the Sahara Desert that has never been charted before. One day, he finds a cave. He carefully goes inside, knowing that snakes and other unpleasant species are known to go in these caves. He is amazed to find a series of images on the wall of the cave. They look like the kinds of caveman images found all over the world, but with added hieroglyphics, like those used in ancient Egypt. However, the cave is on the other side of the Sahara Desert from Egypt, and there are no known or documented instances of humans ever living within five hundred miles of this cave. There are no other examples of these kinds of cave paintings anywhere in that region.

The explorer is puzzled, then starts recording his thoughts in his notes. He concludes that the images must be the result of random snake or insect movements since, from his understanding, humans have never been there.

Does this seem like a sensible logical conclusion?

Of course it isn't logical! Anyone who sees art or writing would conclude that an intelligent, human-like organism created it.

Why is this necessarily so?

There are billions upon billions of examples of art or writing, but they have only ever been created spontaneously by humans on this planet. Random processes might create beautiful images or patterns, even complex ones, but they do not create representations of scenes, ideas or concepts. It is, therefore, entirely consistent with all knowledge to conclude that the paintings were human. To do otherwise is completely illogical.

Now at this point, I can imagine an atheist groaning and muttering that this is just a variant of the "watchmaker argument." Many, including Isaac Newton and Rene Descartes, have invoked the watchmaker argument over the years. William Paley was the most famous proponent:

> *"In crossing a heath, suppose I pitched my foot against a stone, and were asked how the stone came to be there; I might possibly answer, that, for anything I knew to the contrary, it had lain there forever: nor would it perhaps be very easy to show the absurdity of this answer. But suppose I had found a watch upon the ground, and it should be inquired how the watch happened to be in that place; I should hardly think of the answer I had before given, that for anything I knew, the watch might have always been there. ... There must have existed, at some time, and at some place or other, an artificer or artificers, who formed [the watch] for the purpose which we find it actually to answer; who comprehended its construction, and designed its use. ... Every indication of contrivance, every manifestation of design, which existed in the watch, exists in the works of nature; with the difference, on the side of nature, of being greater or more, and that in a degree which exceeds all computation."*
>
> — *William Paley, Natural Theology (1802)*

The watchmaker argument lies at the heart of Intelligent Design. Observing the complexity and elements of apparent design of organs such as the eye and not attributing them to a creator is akin to finding a watch and not assuming it was made by a watchmaker. I know there is much more to modern ID than this, but in essence, this is the central thesis.

One of the many issues I have with using this argument myself is that, while it is true that many biological constructs have the appearance of design, creating biological organs from scratch is not a unique human activity. In fact, it is not something associated with any known intelligent beings at all. There is no precedent. Intelligent humans are able to design mechanical constructs with artificial components, such as a watch, but it does not necessarily follow that only intelligence, human or otherwise, could generate mechanical constructs with biological components. It has a teeny whiff of conflation, but not much. There is a strong inference, based on our intuitive understanding, that makes it plausible to believe these constructs were designed by intelligence, but there is no direct established link between the appearance of post LUCA biological design and intelligence. To me, the watchmaker's analogy, while good, lacks the kind of precision I prefer when arguing about science.

With DNA, we have an entirely different situation. Unlike biological machines, which have never been created from scratch by humans[87], information in the form of codes or language has only ever been created by intelligence. The analogy I used above--namely, of finding paintings and writing--is much more precise than the watch and the watchmaker analogy, due to context.

In my view, the context, or medium in which mechanical design appears has relevance, i.e. biological vs. artificial, and therefore the conclusions drawn can be less absolute.

However, the context or medium that contains information, especially in codes or language, is always irrelevant. Information is only related to the medium in which it appears, in so much as the deliberate organization of that medium is relevant. The words "DNA is cool" convey the same information whether they are written in ink or rocks or dog poop. This is what we discussed in the last section. As a reminder, information is only created by intelligence. Complexity of natural systems only becomes information in the correct understanding of that word when it is observed by a mind. Transmitting information in codes or language are exclusively the domain of intelligence. As we have seen, DNA contains information about other systems in a specific code, which is transmitted by the translation machine. Yes, it is just chemical

---

87   We do now, but only by borrowing from nature.

interactions, just like ink is just ink, but these interactions are organized to transmit specific abstract information in the form of a code, just like ink can be organized to transmit abstract information in the form of language.

This is why the cave picture is an extremely accurate analogy. The context of the painting, a location where no humans are known to have lived, might cause you to look twice to check that it really is a painting, but once it is established that it is indeed a painting, it requires an act of extreme cognitive dissonance to conclude that the picture was created by anything other than a human.

Are there any examples, other than DNA, of information appearing in codes outside of the context of intelligent creation?

No. There are millions of human codes, and possibly millions more animal codes where information is communicated through noises or visual signals by an intelligent brain. The only example where we are not able to physically prove that a code has an intelligent source is DNA, but given our knowledge of information and codes, and that there is a direct unique link between codes and intelligence, regardless of the medium, the only rational conclusion to draw is that the DNA code was generated by intelligence. To do otherwise is no different from concluding that the cave painting and hieroglyphics was drawn by insects.

We can now update our table. Note, I only give a score of 2 for intelligence. If I had allowed for design arguments, it would be much higher, but I have chosen not to. I wish to only include evidence that is directly measurable, rather than inferred. Therefore. I believe that 2 is a fair number.

EVIDENCE	RANDOM PROCESSES		INTELLIGENT INITIATION	
	FOR RANDOM PROCESSES	AGAINST RANDOM PROCESSES	FOR INTELLIGENT INITIATION	AGAINST INTELLIGENT INITIATION
AMOUNT (0-10)	0.5	9	2	0

While there are only two points in the column for intelligent initiation, the *balance* of evidence is nonetheless strongly in favor of intelligence being the source of DNA. That is because, not only is there no real evidence to support the idea that the only other competing explanation, random processes, was the source, but also there is a large amount of evidence against random processes producing the DNA code. Given that there is some evidence for intelligence, the overwhelming balance is, therefore, in favor of intelligence.

Is there a precedent, based on this level of evidence, for generating a more definitive statement to summarize all that we have discussed?

Yes, there is, and to do this I am going to use a statement made in the U.S. Global Research Program in Climate Science Special Report from November 2017 that was accepted globally by other scientists, given the evidence available:

"This assessment concludes, based on extensive evidence, that it is extremely likely that human activities, especially emissions of greenhouse gases, are the dominant cause of the observed warming since the mid-20th century," according to the Climate Science Special Report. **For the warming over the last century, there is no convincing alternative explanation supported by the extent of the observational evidence."**

I am not going to get into the controversy of whether there could be other causes to climate change. In my view, we only have one home, Earth, and we can't take the risk of being wrong on this, so we should err on the side of caution. Even if there was only a 30% chance that warming was due to man-made CO_2, I still believe we should do all we can to produce energy renewably. Let's just get on with it so that we have at least tried to do our best for future generations, and in the process weaned ourselves off an ever-depleting supply of fossil fuels. It's a no-brainer. That little rant aside, my use of this statement as a precedent was because it uses the absence of evidence for alternative causes to conclude that the only cause with some evidence--namely, manmade emissions--is the "dominant" cause.

In other words, the scientific community believes it is acceptable to make a global statement on the causes of climate science on a table like this. (I am not saying this reflects the real evidence, just that this would be the kind of table that the climate panel might have created if they created one.):

	NATURAL CAUSES OF GLOBAL WARMING		MAN-MADE CAUSES OF GLOBAL WARMING	
EVIDENCE	FOR NATURAL CAUSES	AGAINST NATURAL CAUSES	FOR MAN-MADE CAUSES	AGAINST MAN-MADE CAUSES
AMOUNT (0-10)	0	0	3	0

Given this is scientifically acceptable, and given an objective analysis of the facts relating to evidence for and against random causes vs. intelligence, summarized in the table, I believe it is possible to generate a similar statement that best reflects the state of play with regard to the subject at hand. From this statement, we will be able to determine the guilt or innocence of the establishment with regard to the accusation that it has polluted the truth.

Let's quickly remind ourselves of all that we have learned.

8. SUMMARY AND CONCLUSION

To create an elevator statement like the one I promised, and to judge this case, we need to quickly review what we have discussed.

THE CURRENT ESTABLISHMENT ORTHODOXY IS NOT OBJECTIVE

The establishment, through a gradual but acrimonious process, has transformed from being religious-based, and hostile to anyone who challenged the religious orthodoxy, to one that is secular, even atheist, and increasingly hostile to anyone who openly advocates a non-materialistic explanation for the origin of life, among other things. This is evidenced by the behavior of the Canadian establishment figures, Payette and Trudeau, who are part of a wider Western establishment. Their position is based on historical enmity and is not objective. If they are wrong, then the establishment needs to correct course (or be corrected).

GETTING TO THE STARTING LINE

When we looked at the difficulties associated with getting to the starting line, the evidence suggested, from our knowledge of chemistry, that it was extremely unlikely that sufficient quantities of pure biological components would be present for biologically functional molecules to form by random processes. We also learned from our knowledge of chirality that random processes would produce a mix of L and D components but, given that proteins are exclusively L and DNA exclusively D, this presents a very significant problem with no apparent solution. Lastly, we learned that there is no plausible explanation of how a coded cell wall appeared by random processes, and that the oily bubbles or holes in rocks would prove inadequate and remove evolutionary pressure for the generation of such a coded structure.

Even getting to the starting line is extremely unlikely.

THE NUMBERS PROBLEM

We came to understand that the DNA/protein code and translation system is a specific biochemical destination and, due its complexity, could not have appeared *de novo* by random processes. We also learned that, because our current system contains proteins, proteins had to appear at some point. There are no theories or evidence that support the existence of functional collections of "mini-proteins." Therefore, we assumed that the simplest protein that could form as part of the path to LUCA is the simplest fully functional protein that we currently know of, which is about 150 amino acids long. Using one of many similar numbers generated over the years, we calculated that for this protein to appear from random processes would require 10^{45} universes like the one we live in to focus on nothing else but making 150-amino-acid-long proteins every jiffy since the beginning of time, using an infinite supply of pure amino acids.

In addition, we learned that none of the existing theories that try to account for the appearance of life on Earth via random processes have proposed a viable way in which proteins could have formed.

The numbers are too big.

THE CHICKEN AND EGG PROBLEM

Here we came to understand the paradox that lies at the heart of the origin of life conundrum. Which came first--the protein that is coded for by the DNA, or the DNA that is translated by the protein? We uncovered a disturbing fact. High school textbooks suggest that RNA is able to translate DNA without proteins, opening the door to the RNA world theory. We showed that this is a false representation of reality, and that ARS proteins are solely responsible for performing the actual translation step. Therefore, the chicken and egg paradox remained intact and a very strong piece of evidence against the case for random causes.

We had a quick look at the RNA world theory and showed that the word "theory" is a very generous description. It doesn't say how RNA appeared, it doesn't show how proteins appeared, it is unable to account for the appearance of a nucleoside code for amino acids, it doesn't show how proteins took over from ribozymes, and it doesn't show how it transitions from an RNA to a DNA code.

The only rottenness of the chicken and egg problem was the smell coming from the RNA world brain fart.

THE FROZEN ACCIDENT

We established beyond reasonable doubt that DNA is a code by just about any definition that we have for a code. We learned that this code has remained largely unchanged since LUCA, and that the explanation for this is that combinations of more than 2-3 beneficial mutations in the DNA code in one replication cycle are extremely rare. This is evidenced by the scientific knowledge that even the simplest expansion of the code from twenty to twenty-one amino acids, may only have ever occurred once. But whether it was actually an expansion is disputed. We then examined the insurmountable statistical hurdles that would result from upgrading from a doublet system to a triplet system. We concluded that random causes would be extremely unlikely to be able to generate a triplet coding system.

The DNA code is frozen. However, at this point it is starting to look like the opposite of an accident.

THE WHY QUESTION

We looked at the abstract nature of the code and information and the fact that DNA is a code containing abstract information and concluded that random processes could not account for the appearance of this. There is no reason WHY it would happen by random processes, let alone how.

EVIDENCE FOR INTELLIGENCE

Lastly, we established that, due to the fact that abstract codes containing specific information had only ever been observed to be the result of intelligence, combined (not conflated!) with the fact that DNA molecules contain abstract information in an arbitrary code, there is clear, direct, measurable evidence that DNA was the result of intelligence.

CREATING THE STATEMENT

	RANDOM PROCESSES		INTELLIGENT INITIATION	
EVIDENCE	FOR RANDOM PROCESSES	AGAINST RANDOM PROCESSES	FOR INTELLIGENT INITIATION	AGAINST INTELLIGENT INITIATION
AMOUNT (0-10)	0.5	9	2	0

From all of this we were able to systematically, and I believe, objectively, generate a table representing the balance of evidence on this topic which shows:

- There is no physical evidence or plausible theory that suggests random natural processes generated the DNA code.
- There is considerable evidence against the belief that random processes could have generated the DNA code.

Given that this state of understanding, which has persisted for more than fifty years in spite of great advancement in the knowledge of the system, it is scientifically rational to conclude that it is highly unlikely that the DNA code was the result of random natural processes. In the absence of any other evidence, the answer to this problem could be accurately described as unknowable, unsolvable and inconceivable.

However, as we can see from the table, there is other evidence.

- There is direct, measurable evidence that the coded information contained within a DNA molecule was extremely unlikely to be the result of anything other than intelligence.

Given the precedent set by the U.S. Global Research Program in Climate Science, we can condense this into the following balanced elevator statement. No matter what your belief, it is objective and truthful to say that this statement represents the balance of evidence from the sum of current extensive scientific knowledge on the subject of DNA:

There is no evidence or plausible theory supporting the idea that the DNA/ protein coding and translation system was generated by random processes in our universe, and there is considerable evidence against such a belief. Moreover, DNA contains specific information in the form of a code. All the millions of other known examples of coded information are the result of intelligence. Therefore, it is rational to conclude that the most likely source of the DNA code was also intelligence.

This statement, which is balanced and accurate, shows very clearly that the Canadian Governor General was foolish to sneer at those who believe in anything other than random causes. It also shows that Trudeau was unwise to infer that she was speaking the truth, and the truth is in fact the opposite of what she was insinuating. Moreover, the wider establishment, particularly those in education and the media are guilty of misleading the public and students on this most important of topics.

They are all guilty of polluting a scared truth, a truth that points to a completely different understanding of our reality than the one we are currently being pushed towards believing. At its heart is the question of eternity. The current establishment's materialistic orthodoxy removes any contemplation of the possibility of eternity, but the truth reveals that maybe we should take a second look.

Before we part company, I will expand on that a bit more.

9. FINAL THOUGHTS

9.1 FOR ATHEISTS

First of all, congratulations for getting this far, it can't have been easy. I can imagine there would have been much bucking and snorting along the way, much like one of my editors who peppered the manuscript with comments like "But it's not really impossible" or "but you haven't really proven this" or "but science could still find the answer to this". To a point, those comments are derived from valid observations. The evidence proves nothing. Short of intelligence posting a selfie of the moment that it created DNA on Instagram, we will never have definitive proof. Also, yes, a chance of 1 in 10^{164} is not impossible from a purely technical perspective, and of course the science of the gaps (one day science will find the answer) is always available as a mental emergency exit. It is always a possibility that a future genius will solve this puzzle. I know that no matter how persuasive the evidence that is provided, you will most likely fall back on this. Remember the opening remarks? However, in one last plea to you, I ask you to consider this quote from John Locke's *Essay Concerning Human Understanding*:

"Let us then suppose the mind to be, as we say, white paper void of all characters, without any ideas. How comes it to be furnished? Whence comes it by that vast store which the busy and boundless fancy of man has painted on it with an almost endless variety? Whence has it all the materials of reason and knowledge? To this I answer, in one word, from EXPERIENCE."

Imagine if you came to the evidence presented in this book with a completely blank mind, with no preformed prejudices, or positions. Just on the face of the evidence presented here, would it still make sense to believe that life was the result of Random Processes?

Anyway, whether that makes you think twice or not, there is one thing that I ask of all atheists who have got this far - if you have been rude to or prejudiced against people of faith, please stop right now. Even if you can't accept that the evidence doesn't favor intelligent initiation, you must accept the findings of your co-atheists who say there has been little to zero scientific progress on this subject, and therefore your subjective belief that life was the result of random processes is *at least equally* a position of faith as those you may have been ridiculing. I would actually argue that it is a position that requires more faith.

9.2 AGNOSTICS/FENCE SITTERS

I suspect that you will have asked many of the questions that atheists would have asked, but you will have also been troubled by the strength of evidence against what the establishment has been telling you all these years--namely, that life was the result of random processes. But it is blue-pill red-pill time... get off the fence.

For millennials and younger, out of all the people living in Western countries: I suspect you are the ones who would have had the least exposure to non-establishment ideas on the origin of life, especially during your education. If the conclusions derived in this book are even half true, then you have been lied to since you were born. Not only were you told a little lie, you were told a whopping, great fat lie about a central truth – where we come from, and who we are. How does that make you feel?

For Gen Xers, the ones between the boomers and the millennials, and who seem to have avoided the media-driven, intergenerational insults that have been stirring up a creeping rage, and Boomers, the ones the rage is focused on: you will both have had exposure to at least some religious teaching over the years, so if you are agnostic there are other reasons than just your education. Media, disappointment, lack of any religious experience--for whatever reason, you have probably chosen to shelve the idea of an intelligent creator, as the evidence to support such a notion does not exist. Or so you thought. Now that you have read this book, what do you think? Do you feel challenged? Do you feel there might be more reason than before to investigate philosophical and religious teachings and texts? If so, then get off the fence and get on with the search.

I have not discussed my Christian beliefs in this text but will do so in the last section addressed to those who share my particular belief. (And if you aren't a believer, you may find some of that a bit "weird") The truth is that none of what has been presented here allows us to identify who or what that intelligence is.

I have other reasons for my faith in Jesus Christ being "The One", this data only supports the belief that life was created, which feeds into my overall faith. This data could just as easily support the idea that life is actually a computer simulation, or that a "God" made us, but rather than being the loving God that I know through Jesus, he is some tyrannical monster who takes pleasure in our suffering. These ideas all rely on an intelligence that created the Universe, or illusion of a Universe. The data could also support the belief that we were created by an alien intelligence from within our universe. However, this is just an advanced version of panspermia, moving the problem to somewhere else in our universe and therefore does not solve the question of how life came into being in our universe.

If you have been shaken up by this book and feel there is truth in what has been said, then start your search for *the* truth. Try not to let your preconceptions cloud your view; be open to the idea that the answer has been right in front of you all these years, but also look into all religions, gain an understanding of these things. Also, if this book is successful enough, then I will write a follow up book on the nature of consciousness from a scientific perspective, a subject that is actually of more interest to me than the origin of life, and one that I have been reading about and researching to an even greater extent. If you think that might be something worth reading, then please tell your friends about this book as that will encourage (and finance) me to work hard on producing something thought-provoking.

9.3 LAST THOUGHTS FOR CHRISTIANS

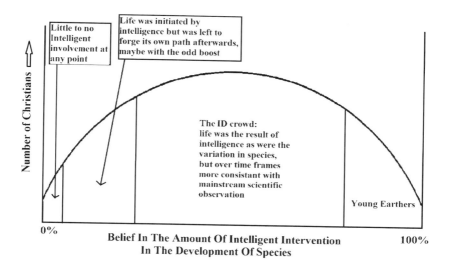

There is a spectrum of belief amongst Christians in this area of apologetics that ranges from theistic evolution, including allowing for random processes forming life, through to six-day creation and the belief in a young Earth. All but those on the far left of the spectrum believe in an interventionist God. Theistic evolutionists, such as those at BioLogos, suggest that we Christians shouldn't ever fill unexplained scientific phenomenon with the words "God did it." In particular, they caution against using this "God of the Gaps" argument in the origins of life context. They argue that the church has been proven wrong so many times, that if we make a stand against random natural processes on the issue of the origin of life, then we will only be proven wrong again as science will inevitably provide the answer.

Firstly, the past cannot always be relied on as a predictor of the future. Just because science has found answers in the past on other questions, does not mean that it is destined to do so on this subject in the future. If anything, science has proven utterly inadequate on this topic, despite its best efforts.

Secondly, the science of DNA and translation is actually very well understood and, as I have stated, many scientists now are conceding that the answer is unknowable or unsolvable; we will never be able to answer this problem. They recognize that the statistical and conceptual hurdles are so immense that it is unlikely a plausible theory will ever be generated. However, we know more than that; we know that the evidence actually shows that it is not statistically or conceptually possible for life to have appeared by random processes. This is much more than a lack of evidence; this is evidence against life being generated by random processes.

Thirdly, this is not a God of the Gaps argument. It is an argument based on a scientifically quantifiable piece of evidence to support intelligent involvement in the beginning of life. It is not just relying on the absence of other evidence or other theories; it actually provides measurable, direct scientific evidence in support of the argument. Therefore, I strongly rebuke the approach of BioLogos[88].

I believe the rest of us have more that unites us than divides us, even if it doesn't feel like that at times. The vast majority of true Christians, believe that at the very least, the appearance of life on Earth was the specific intentional act of a creator God. It wasn't just the result of some laws God put in place at the beginning of the universe; it was a specific intervention in the history of Earth (and possibly many other planets, too). This is so fundamental to our understanding of who God is that I strongly believe we should unite around this one topic when it comes to apologetics. If God did not, at the

---

88   The more I look at Biologos, the less I trust them. They make statements of faith that seem aligned to core Christian beliefs, but the articles I have read on their site regarding the origins of life appear to go out of their way to discount any possibility that life was the result of direct intelligent intervention outside of natural laws. Not only does this appear to contradict the most liberal interpretation of scripture, it actually goes against the scientific evidence. The Apostles warn us about false teachers, and I personally would put Biologos in this group.

very least, create life, then the God of the Bible does not exist; he is something else entirely. He is not a personal God who hears our prayers and is intimate with the details of our lives. Instead, he is either not there, or some distant, disinterested being that is merely a curious observer. Jesus was just another prophet, and not the specific intervention of God in history to rescue us from sin. There is a precedent in the Bible for making such a stand:

"If there is no resurrection of the dead, then not even Christ has been raised. And if Christ has not been raised, our preaching is useless and so is your faith."

– *1 Corinthians 15:13-14 NIV*

There are a number of Christian beliefs that go against establishment groupthink that we must make a stand on, or our faith is "useless." The resurrection is one[89], and I believe the rational belief that God created life is another.

I personally believe that, whether we like it or not, we don't have any choice but to fight this issue. The good news, as I have said repeatedly, and demonstrated in this book, is that the science is strongly on our side.

This cannot be said with such confidence about other areas of scientific apologetics. Now I know that everyone has their favorite tidbit of information that supports their particular place on the spectrum of belief with regards to creation, but I ask you to "car park" that for the moment,

---

89   As an aside, and a prelude to things I will discuss in future books, I wonder if "miracles" such as the resurrection, healings, the virgin birth etc. may be explained by a completely unproven and unsubstantiated idea that God interacts with this world through quantum manipulations. Of course, as I mentioned at the beginning, I currently lack an in-depth understanding of quantum mechanics, and accept that such a thought may be complete nonsense - although no more so than physicists believing undirected quantum processes could generate information in the form of a code. It is something I intend to learn more about to enable me to elucidate further.

especially in discussion with people outside of the church. I also ask people who are more towards the left of the spectrum, who perhaps believe in an old Earth, to stop making snarky remarks about those whose belief in the Bible being the literal word of God causes them to stand by six-day creation. We all, except those on the far left of the spectrum, believe in an awesome God who created life...whether he did it on the first of six days, or the first of a trillion days.

My advice for the moment is the same for all discussions on the events after the creation of life. Whether you believe God was involved with the formation of every being that ever existed, or just got the ball rolling with the first cell and then stood back and watched, we all still believe that God created that first cell and we should unite in strongly voicing our position on this. God is the creator and author of life either way.

I believe that our ultimate goal as a church on the subject of creation apologetics should be this, and this only:

To level the playing field.

Nothing else. Just to neutralize a corrosive and insidious, atheist, establishment lie, propagated by a lazy media and complicit educational system. In my view, it currently presents one of the most significant barriers to belief since it kills any nascent curiosity about God.

We should use language that is balanced and reflects the scientific evidence. On this subject, our goal should not be to convince others that science shows God created life, but instead to show that, at the very least, it is as rational to believe that life was created by an intelligent being as it was by random processes and, if anything, the balance of evidence favors belief in an intelligent initiator. Therefore, having belief in a creator God is not at odds with our rational understanding of science. This stance opens the door to discussions on other evidence about God and who that God might be, and thereby creates opportunities to bring others into relationship with our Savior.

My side of the family argued passionately for our creator at the dinner table against Charles Darwin all those years ago, and for a long time it has felt like they had lost. My hope is that this book will be a part of turning the tide. To do that requires what I have just asked, for Christians to unite on this subject[90].

90   And recommend this book to their friends!

APPENDIX

The aim of the appendix is to go into a bit more detail on some topics that might have proved a distraction for those who don't want to get into the weeds in the main book. However, even though I will address some topics in detail, especially the RNA world, there are other areas for which I will just highlight a few pertinent examples. This is particularly the case for the "Getting to the Starting Line" section. I will try to provide additional references where appropriate. What this is not, though, is an exhaustive analysis and debunking of every single theory that has been floated. The main text addresses the key elements of showing that there is no evidence for random causes, that the theories to date are not viable and that the evidence for intelligence is real. What is added here are a few items to provide a bit more detail. Also, I abandon any attempts at not being too scientific or dry. This is just for your information, not entertainment.

APPENDIX SECTIONS

SECTION 1: ABIOGENESIS AND THE SPONTANEOUS APPEARANCE OF STARTING MOLECULES BY RANDOM PROCESSES

But if (& oh what a big if) we could conceive of some warm little pond with all of ammonia & phosphoric salts, -light, heat electricity present, that a protein compound was chemically formed, ready to undergo still more complex changes...

-- Charles Darwin, from an archive of his private correspondence.

In the main text, I focus on the broader problems facing the production of suitable starting materials--namely, the idea that no matter what kind of chemistry took place in an open early Earth environment, it would not have been able to produce sufficient quantities of suitable starting materials in adequate purity. This applies to any theory that has been developed that proposes a reaction pathway by which amino acids or nucleosides may have formed in prebiotic conditions.

My key objection is not that amino acids or nucleosides ever formed, but that if they ever did form, it would have been in inadequate quantities and riddled with impurities. This was the intention of the cake analogy. To me, it is a key factor in presenting an insurmountable barrier to large functional biomolecules ever forming. However, any text discussing the origin of life would be incomplete if it didn't, at the very least, mention the Miller Urey experiment, and one or two of the more recent iterations.

OPARIN-HALDANE HYPOTHESIS

During the 1920s, Russian scientist, Aleksandr Oparin (a Russian communist and, therefore, someone with a strong vested interest in producing theories that championed atheism), and English Scientist, John Haldane, also a communist atheist, generated a "broad strokes" theory of how life might first have appeared from very basic inorganic chemicals.

They proposed that the early Earth had a strong reducing atmosphere (something that has now been shown to be unlikely) and that simple inorganic molecules reacted using energy from the sun or lightning to produce amino acids or nucleotides, which may have accumulated in the oceans, creating a primordial soup.

These building blocks then combined further, creating complex polymers like proteins and oligonucleotides, perhaps in places like the water's edge where there were greater concentrations of the materials and where evaporation had occurred. These polymers then formed structures that were capable of sustaining or replicating themselves. This could either have occurred through clusters that carried out metabolism (Oparin proposed this) or that became enclosed in membranes to make cell-like structures (Haldane's suggestion).

Given the lack of knowledge of some of the details of cellular and biochemical structure, the theory was light on details and formed a framework rather than a detailed thesis. This framework has, to some extent, stayed intact to this day when origin of life researchers attempt to generate new theories on the appearance of organic molecular building blocks.

MILLER UREY

This famous experiment adds credence to the Oparin and Haldane framework, albeit a good thirty years later. I am going to quote a summary of the experiment directly from the Wikipedia page as of June 2018:

"The Miller–Urey experiment[91] (or Miller experiment)[92] was a chemical experiment that simulated the conditions thought at the time to be present on the early Earth, and tested the chemical origin of life under those conditions. The experiment supported Alexander Oparin's and J. B. S. Haldane's hypothesis that putative conditions on the primitive Earth favoured chemical reactions that synthesized more complex organic compounds from simpler inorganic precursors. Considered to be the classic experiment investigating abiogenesis, it was conducted in 1952[93] by Stanley Miller, with assistance from Harold Urey, at the University of Chicago and later the University of California, San Diego and published the following year.[94],[95],[96]

"After Miller's death in 2007, scientists examining sealed vials preserved from the original experiments were able to show that there were actually well over 20 different amino acids produced in Miller's original experiments. That is considerably more than what Miller originally reported, and more than the 20

---

91   Hill HG, Nuth JA (2003). «The catalytic potential of cosmic dust: implications for prebiotic chemistry in the solar nebula and other protoplanetary systems». Astrobiology. 3 (2): 291–304. .

92   Balm SP; Hare J.P.; Kroto HW (1991). "The analysis of comet mass spectrometric data". Space Science Reviews. 56: 185–9.

93   Bada, Jeffrey L. (2000). "Stanley Miller's 70th Birthday" (PDF). Origins of Life and Evolution of the Biosphere. Netherlands: Kluwer Academic Publishers. 30: 107–12.

94   Miller, Stanley L. (1953). "Production of Amino Acids Under Possible Primitive Earth Conditions" (PDF). Science. 117 (3046): 528–9.

95   Miller, Stanley L.; Harold C. Urey (1959). "Organic Compound Synthesis on the Primitive Earth". Science. 130 (3370): 245–51

96   A. Lazcano; J. L. Bada (2004). "The 1953 Stanley L. Miller Experiment: Fifty Years of Prebiotic Organic Chemistry". Origins of Life and Evolution of Biospheres. 33 (3): 235–242

that naturally occur in life. More recent evidence suggests that Earth's original atmosphere might have had a composition different from the gas used in the Miller experiment. But prebiotic experiments continue to produce racemic mixtures of simple to complex compounds under varying conditions.[97]"

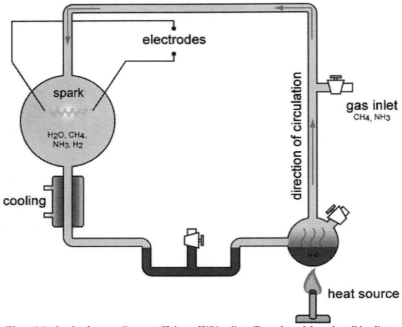

By The original uploader was Carny at Hebrew Wikipedia. - Transferred from he.wikipedia to Commons., CC BY 2.5, https://commons.wikimedia.org/w/index.php?curid=2173230

There are a couple of things to note about this. The first is that the Wikipedia account correctly states that the experiment may not have reproduced the early Earth atmosphere correctly, and that recent evidence points to an atmosphere that wasn't reducing and was thereby unlikely to support the spontaneous appearance of amino acids via this route.

The lack of certainty about the composition of the early Earth environment, and the entire lack of any evidence at all that amino acids appeared in

---

97   Bada, Jeffrey L. (2013). "New insights into prebiotic chemistry from Stanley Miller's spark discharge experiments". *Chemical Society Reviews*. **42** (5): 2186–96

meaningful quantities due to the expanse of time, are central problems facing all researchers who try to develop solutions to this problem. That is why there is only 0.5 in the "Evidence For" column; there is virtually no actual evidence to support that this first step ever happened.

The second thing to note is the picture. This "simulation" was conducted in a lab using controlled conditions with pure reagents. Now if there had been a reducing atmosphere, there may well have been moments when just such a perfect setup occurred and small quantities of amino acids formed, but there would have been other by-products of the reaction present, as well. The presence of other impurities would have made the likelihood of even a small polypeptide chain forming extremely remote, since they would cross-react with the pure product.

Then there is stereochemistry. The mixture produced in the Miller Urey experiment was racemic, and we know that all amino acids used in proteins are homochiral, and L. This is a problem facing virtually all of these types of proposed solutions, and no truly adequate explanation has been provided which would overcome this significant hurdle.

MORE RECENT THEORIES

John D. Sutherland, who is a researcher at Cambridge University in the U.K., and one of the more "respected" origin of life researchers currently touting theories, has proposed a "unifying" pathway[98], or set of pathways, that relies on concurrent reactions, under similar conditions and using similar starting materials (hydrogen cyanide, hydrogen sulfide, phosphates, and water), some of which would need to be delivered by meteorites. These concurrent reactions occur in different streams flowing down mountains, generating building blocks

---

98   Studies on the origin of life – the end of the beginning; J.D. Sutherland; Nature Reviews Chemistry, Vol. 1:12 (2017)

which then combine through a confluence of streams, or by evaporation and rehydration cycles. Through this approach, Sutherland and coworkers have produced what they describe as "plausible chemical pathways" that could produce both amino acids and nucleotides, which could then go on to form the polymers central to life's processes. They've even suggested that a cell could spontaneously form in a matter of minutes from this type of process!

Firstly, I must point out that their chemistry is technically correct; the complex multistep pathways they propose could indeed produce a number of the basic biomolecular building blocks required to generate DNA or RNA and proteins. However, the "theory" suffers from the same problems that I describe in the *Getting Started* section. None of these processes would produce 100% yields of pure starting materials; as with most chemical reactions, there would be by-products, and that is when the conditions are perfect (i.e. right temperature, concentrations, pH etc.). Now Sutherland suggests that these by-products would also have utility in the process of forming biomolecules, but rather than cross-react with the main product, they are washed away by the stream in which they are flowing. Moreover, less favorable by-products would be destroyed by radiation. A similar set of perfectly advantageous circumstances is happening in a nearby stream, producing another step in the reaction pathway towards an amino acid or nucleoside and, similarly, the unwanted by-products are being washed away or destroyed by UV radiation. Fortuitously, the required pure products unite at the bottom of the hill (not with the by-products, though). A dry spell concentrates them and causes them to react together and generate the important biomolecule. It's Miller-Urey on steroids.

Ultimately, there is no evidence that this did happen, and while the chemical routes might be plausible in a lab using pure starting materials and controlled conditions, the chance of it happening in the way described are vanishingly small and require just a little bit too much luck (before you need a load more for the numbers problem). Then, of course, this has to happen over and over, and once again produce the right stereochemistry, and produce trillions of tons of material for there to be sufficient amounts to be able to form a functional protein or RNA molecule. No natural environment is conducive to producing biological starting materials on an industrial scale.

Ramanarayanan Krishnamurthy from the SCRIPPS Institute has also made contributions in this field, as have numerous others. *Nature* is filled with articles proposing new ideas as to how these important starting materials could have appeared. They cite chemical pathways that have been generated in the lab and that support the validity of these ideas.

One day, and it may even be Sutherland's idea, a scientist, most likely a chemist, will produce a really robust chemical route to the required starting materials. I have that much confidence in the abilities of my fellow scientists. I remember when I was an undergraduate and was discussing LCD screen technology with a physicist who was working on LCD crystals. At the time, the screens that had been produced were only in greyscale and suffered from poor refresh rates and ghosting. I said that one day they would figure out how to create color LCD screens with refresh rates and resolutions on a par with TVs. He laughed in my face. So I am not one to underestimate the ability of scientists and, as I say, it may well be that Sutherland's is the best hypothesis as to how biological starting materials might have appeared; there might be better ones generated over the years, but even if there are, they will all suffer from two key problems:

1. There is no way of proving them. Seems unfair to keep dragging that one out, but it is nonetheless true. The geological evidence will not be available. The lack of evidence to prove or help formulate a theory is one of the reasons why there are so many different proposed routes. This has advantages and disadvantages. The advantage is that chemists have a relatively blank canvas with which to work. The disadvantage is the resulting multiple theories and the apparent lack of any conclusive consensus, adding to the general sense of lots of people going nowhere fast.

2. While viable in a lab, under the direction of a trained chemist and using pure materials and perfectly adjusted conditions, they will never be able to validate the viability of these processes in any real-world environment. Moreover, even if these routes were potentially viable, they would not be able to produce the required amounts of pure starting material to generate functional biomolecules like RNA or proteins. That is because "natural" chemistry is messy, unless it is only making extremely simple inorganic molecules, or under the direction of proteins.

To be fair to Sutherland, he does for the most part recognize the enormity of the problem and the lack of progress we have made when it comes to providing answers to the questions surrounding the origin of life. He says this in the introduction to his 2017 review:

"Understanding how life on Earth might have originated is the major goal of origins of life chemistry. To proceed from simple feedstock molecules and energy sources to a living system requires extensive synthesis and coordinated assembly to occur over numerous steps, which are governed only by environmental factors and inherent chemical reactivity. Demonstrating such a process in the laboratory would show how life can start from the inanimate. If the starting materials were irrefutably primordial and the end result happened to bear an uncanny resemblance to extant biology — for what turned out to be purely chemical reasons, albeit elegantly subtle ones — then it could be a recapitulation of the way that natural life originated. We are not yet close to achieving this end, but recent results suggest that we may have nearly finished the first phase: the beginning."

I would argue, for the reasons stated above that they are nowhere near the end of the beginning even. There is no evidence that his proposed pathway took place, and plenty of evidence--namely, the chemistry-related challenges of undertaking the kind of chemical paths he describes--against it. However, he is even more honest in his concluding remarks:

"The random synthesis of peptides then appears particularly unattractive and very early coded synthesis is suggested. Clearly we are not yet even at the beginning of our quest to understand it, but the end of the beginning is offering up some very tantalizing clues about the origin of life."

He illustrates this in a graph from this paper. I have produced a simplified copy which I used in the main text and also below. The graph is a very good visual illustration of just how little progress has been made with regard to understanding the science behind, or generating theories to explain the arrival of, LUCA. This is why I relegate his work, along with all the other work trying to produce starting materials to Getting to the Starting Line. Most of the other arguments in the first section of this book work on the assumption that vast amounts of starting material are available, which is of course nonsense, but if I don't do that then we can't discuss the biggest barriers. Some would argue that the RNA world fills in lots of the gaps. In Section 7 of the Appendix, I show why that is not the case. Physicists, recognizing that simple chemistry etc. is unable to answer the questions around the origin of life, are now prone to waving their quantum mechanical magic wand at the problem. I will look at this in Section 3...very briefly. One of the key things to note, though, is that most of the serious researchers or theorists in this field openly recognize and reiterate over and over in scientific journals how little progress we have made towards finding answers, which is in stark contrast to the establishment position and media hyperbole.

In the graph below, the arrow is pointing to a part of the line that is above zero, it just doesn't look like it due to scaling. The point that Sutherland is trying to make, and explicitly says, is that we are a long, long way from LUCA.

Transition From Inorganic Starting Materials to First Cell

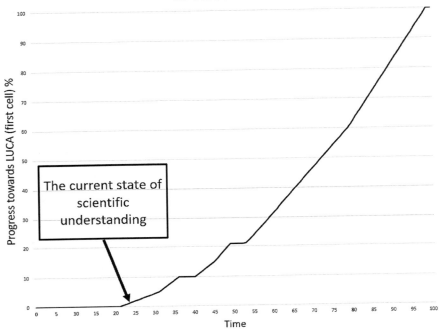

CELL WALL

Below is a picture from Wikipedia (public domain by Ladyofhats - https://en.wikipedia.org/wiki/Cell_wall#/media/File:Plant_cell_wall_diagram-en.svg) of a segment of plant cell wall.

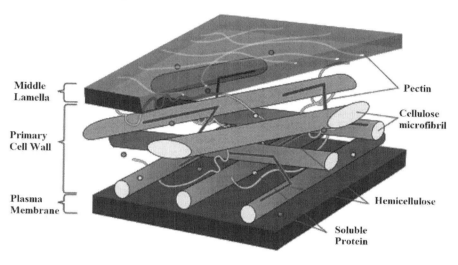

As you can see, it consists of a number of different structures, all of which are coded for by the DNA in the cell. Animal cell walls are even more complex (technically speaking animal cells don't have cell "walls", they have cell membranes constructed from different materials - so I do care about biologists after all! I just wanted to keep things simple in the main text and was prepared to risk a few Biologists having a meltdown in the process). The simplest cell walls are very complex, and even viral cell walls (viruses aren't really cells per se, and their "walls" are called envelopes) are coded for in long sections of their genetic code. Compartmentalization theory does nothing to answer the question of how the first cell wall appeared and was coded for. There are no theories that explain this, and yet it is vital for life to even get going.

SECTION 2: AMINO ACIDS

Below is a table[99] of the chemical structures of the amino acids (from Wikipedia and created by Dan Cojocari, Princess Margaret Cancer Centre, University of Toronto).

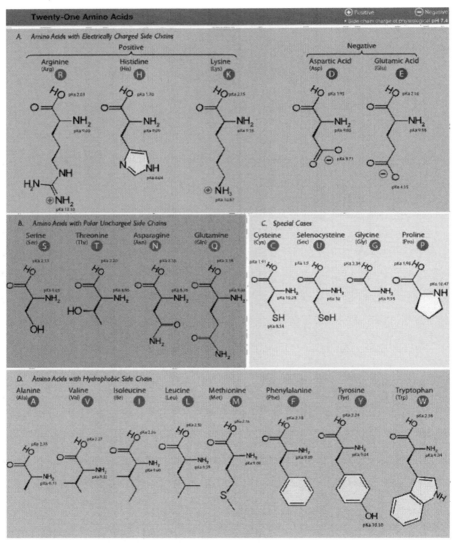

---

99   Source: Dancojocari - https://commons.wikimedia.org/w/index.php?curid=9176441

As you can see, they are all based around the central amino acid motif (the CO_2H group on one end is a carboxylic acid group, and NH_2 is an amino group, hence amino acid).

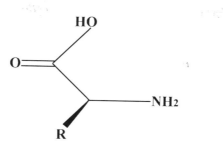

The R group is the moiety that is changed between amino acids and determines the different properties that each amino acid possesses. The simplest amino acid is glycine, and because the R group is in fact H (Hydrogen), it is the only amino acid that is achiral...i.e. no mirror image, or L or D version.

This table also adds the property of hydrophobicity, "fear of water," or in this case water-repelling. The large groups add bulk, which makes them space-taking but also gives them a propensity to push away from water, or alcohol (OH) containing molecules or moieties. All these different properties are what determine the folded shape of the chain of amino acids that makes the protein, and their ability to interact with their surrounding environment and perform various actions.

Notice, there are twenty-one amino acids in this table, with selenocysteine being the additional amino acid. There is, in fact, another one, making twenty-two naturally occurring amino acids. I cover these non-universal amino acids in Section 4. They are basically coded in a slightly different way and exist in a few bacteria.

SECTION 3: THEORIES OTHER THAN THE RNA WORLD

It is a capital mistake to theorize before one has data. Insensibly one begins to twist facts to suit theories, instead of theories to suit facts.

--Sherlock Holmes

As I point out in the main section, we are where we are and we had to get here somehow. You can either work backwards, which no one does, start at the beginning and not get very far, or as with the RNA world, start in the middle of a completely different journey. More on that in Section 7.

In fact, there are two main, competing, "overarching" theories as to how life came into being: genetics first (DNA or RNA) and metabolism first. Protein first is not really considered as a potential candidate anymore, mostly due to the huge numbers problem, and also due to the lack of a mechanism for passing on information, a prerequisite for any living system. As I said, the RNA world theory is the only real genetics first theory. Here we look at metabolism first and some other rather strange ideas.

METABOLISM FIRST

First proposed in 1988[100] by Günter Wächtershäuser, this theory proposes that small molecules such as carbon dioxide and water form slightly more complex molecules like acetate in a thermodynamically favorable location, like a steam vent. These smaller molecules combine to generate more complex molecules that eventually catalyze reactions themselves, and thus create an

---

100   Wächtershäuser G(1988) Before enzymes and templates: Theory of surface metabolism. Microbiol Rev 52:452–484.

auto-catalytic cycle. The energy initially comes from an external source but, as they "evolve," they are able to produce their own energy through a kind of metabolism. These systems, because they are self-regenerating, essentially pass on the "information" from, or about, the previous system and therefore in some way show Darwinian evolution. At some point, this system makes RNA and/or proteins and we arrive at the system we have today. Wächtershäuser's original paper in 1988 runs to thirty-three pages, describing how all this might happen. Have you heard of Wächtershäuser? Exactly. If he'd cracked the origin of life puzzle, he would be a household name.

There is no evidence that such a system did or could exist outside of a living system. More importantly, there is no viable proposed route as to how you would generate the RNA or proteins. Finally, and fatally, the information would not be passed on efficiently enough to meet the requirements of a system displaying Darwinian evolution[101]. The information would dissipate as the system produced multiple forms of molecule and no template in which to store the information. Because the system is not compartmentalized, the bits and pieces would just float off. It is a fantasy system and largely abandoned now, although it does surface from time to time. Like all of these "theories," there is some scientific merit in the idea, but where they really fall down is on how information is created in storable format and passed between generations, and how the whole lot is held together. This is where it is vital to understand that complexity only becomes information when it is observed.

---

101   Lack of evolvability in self-sustaining autocatalytic networks constraints metabolism-first scenarios for the origin of life Vera Vasas, Eörs Szathmáry, and Mauro Santos; PNAS January 26, 2010. 107 (4) 1470-1475; https://doi.org/10.1073/pnas.0912628107

COMPARTMENTALIZATION FIRST

This is not really a stand-alone theory as such, but is thrown in there to try to overcome the issue of a cell wall. Every single theory faces a number of problems, but all face the problem of keeping the stuff together in the same place. If you have reactions going on near a deep sea vent, then the newly formed materials or systems would spread out into the surrounding ocean; they would no longer be systems and no longer be able to work together. If you have polymers etc. forming in clay (see Section 4) then they will be in one place, but the problem then becomes how you move them around without diluting them. At some point, any molecules produced would need to be flushed out with water, where they then would dilute and no longer be in close enough association to act as systems. The oligonucleotides and/or peptides need to be free-moving to be able to associate with other molecules, but within an enclosed system so they don't just swan off.

Various "solutions" have been proposed to overcome this, but they whither quickly under even the lightest of scrutiny. The two solutions that are most commonly touted are small compartments on the surface of minerals, or oily lipid membranes that spontaneously form like bubbles around the systems. The (obvious) problem with both of these is that they are closed systems. If the pocket on a mineral has an opening, for it to act as a compartment, the opening must be so small as to severely hinder the in and out flow of materials, or it is not a compartment. This is its downfall, too, as it is, for all intents and purposes, closed. You can't have your cake and eat it. In either case, they have to "become" closed with just the right quantities and type of ingredients in them; but what then? Whatever system you have inside the compartment, it is going to need to remove waste materials from whatever process has occurred inside, and it will need to acquire new materials to repeat the process. In a closed system, this cannot happen. Also, because there is no interaction with the outside world, how can it pass on the information of the system to create successive generations? It can only be a one-off. But there is a bigger problem still.

If you have a system that works inside a bubble or mineral pocket, then the need for generating a compartment is removed. There is no evolutionary driver for the creation of a cell wall. The chicken and egg paradox of the cell wall remains. The code for a cell wall exists in the genetic code and, at some point, it needed to be generated. Shoving all the stuff in a pre-made compartment removes the driving reason to generate it, yet without it the materials would indeed just wander off.

The compartmentalization first "theories" are embarrassing attempts to hide one of the biggest conundrums and obstacles to a random processes explanation--namely, the appearance of a cell wall. They don't work, and they don't solve the problem.

CLAY THEORIES

The late Graham Cairns-Smith proposed[102] that life first formed on clay, since clay possesses properties that suggest it could pass on information. Even though the theory sounds completely batty, montmorillonite clay has indeed been shown to catalyze the formation of RNA polymers[103] (under strict laboratory conditions). In essence, his theory proposes that, because clay crystals calve away from the parent structure with near identical composition, they could form the basis for some kind of template for passing on genetic information. It is not a theory that is taken to be of much use now. If something is bound

---

102   The origin of life and the nature of the primitive gene, Journal of Theoretical Biology: Volume 10, Issue 1, January 1966, Pages 53-88

103   Montmorillonite-catalysed formation of RNA oligomers: the possible role of catalysis in the origins of life; James P Ferris; Philos Trans R Soc Lond B Biol Sci. 2006 Oct 29; 361(1474): 1777–1786

to the clay, how would it unbind and dissipate? Once it did, where would it go and find other molecules of a similar inclination? How would the molecules move about in the first place inside clay-like structures? Even if all these insurmountable questions were "mountable," how does the information and the sequences of functional molecules get encoded? The molecules forming would just be random; the clay itself does not form an informational template, rather just a physical template.

QUANTUM MECHANICAL EXPLANATIONS

Paul Davies wrote a book on the origin of life. He was (and still is, as far as I am aware) an atheist physicist who looked into the origin of life question expecting to find an easy answer. (Typical hubris of physicists who, maybe rightly, believe they are infinitely more intelligent than anyone else on the planet.) He was humbled by the problem, and wrote about it, coming to many of the conclusions that Stephen Meyer and many others have come to--there is no explanation for the origin of DNA and, specifically, the origin of the code/information in DNA. His conclusion was that chemistry would not be able to explain it but, rather, that physics would, and coined the term "complexity" theory--namely, that there is some as yet undiscovered force of nature that drives systems towards complexity, including the addition of information. He proposed this around the turn of the century. A paper was published in 2017[104], by a colleague of Paul Davies, that shows that no progress has been made to date. The abstract is as follows:

"The origins of life stands among the great open scientific questions of our time. While a number of proposals exist for possible starting points in the pathway from non-living to living matter, these have so far not achieved states

---

104   Origins of life: a problem for physics, a key issues review; Sara Imari Walker 2017 Rep. Prog. Phys. 80 092601

of complexity that are anywhere near that of even the simplest living systems. A key challenge is identifying the properties of living matter that might distinguish living and non-living physical systems such that we might build new life in the lab. This review is geared towards covering major viewpoints on the origin of life for those new to the origin of life field, with a forward look towards considering what it might take for a physical theory that universally explains the phenomenon of life to arise from the seemingly disconnected array of ideas proposed thus far. The hope is that a theory akin to our other theories in fundamental physics might one day emerge to explain the phenomenon of life, and in turn finally permit solving its origins."

This is akin to the analogy of the husband taking over the shelf assembly from his wife and kids and saying "let me do it!" "We are physicists, we have solved all other puzzles (OK, aside from the origin of the universe and consciousness) so let us have a quick look at this and we will have it licked." Of course, I am being facetious, and the article is an excellent review of all the various theories, with some physics formulas thrown in just to ensure that readers will be convinced of the authority of the authors. It is noteworthy that the abstract is very similar in tone to Sutherland's. Basically, "we haven't got a bloody clue." They are honest in stating this, unlike the establishment, who infer that the problem is all but solved. Also, like Sutherland and Davies, Walker points to the deeper problem:

"The idea of information is itself abstract, but it must be the case that each bit of information is instantiated in physical degrees of freedom. "Information is physical!" in the words of Rolf Landauer. Whether fundamental or an epiphenomenon, the causal role of information in biology represents one of the hardest explanatory problems for solving the origins of life."

Although I am being a bit cheeky about their inherent belief in the ability of scientists in their subject area to solve one of the greatest questions we have, a part of me wonders if they are on to something. If, as I do, you believe that the entire universe was the result of an intentional creative act by a being of enormous intelligence and power (or a programmer), then the "tools" by which this being interacts with his/her/its "creation" would be an interface that joins the two realms. This is where quantum mechanics may come into it, and the ability to manipulate quantum states through a "force" or "will" may actually be that mechanism. This is not so different from what Davies proposes, and what Walker is seeking, except for there being an intelligence behind it.

SECTION 4: THE CODE

Below is the standard table of the DNA code. Notice, there is more than one codon for each of the twenty standard amino acids and for the stop signal. This is called redundancy, and I discuss this below.

DNA code (from Wikipedia under the creative commons license) and RNA in the table after that (same source and licensing):

Amino acids biochemical properties: nonpolar | polar | basic | acidic   Termination, stop codon

Standard genetic code

1st base	2nd base T	2nd base C	2nd base A	2nd base G	3rd base
T	TTT (Phe/F) Phenylalanine	TCT	TAT (Tyr/Y) Tyrosine	TGT (Cys/C) Cysteine	T
	TTC	TCC (Ser/S) Serine	TAC	TGC	C
	TTA (Leu/L) Leucine	TCA	TAA[B] Stop (Ochre)	TGA[B] Stop (Opal)	A
	TTG	TCG	TAG[B] Stop (Amber)	TGG (Trp/W) Tryptophan	G
C	CTT (Leu/L) Leucine	CCT	CAT (His/H) Histidine	CGT	T
	CTC	CCC (Pro/P) Proline	CAC	CGC (Arg/R) Arginine	C
	CTA	CCA	CAA (Gln/Q) Glutamine	CGA	A
	CTG	CCG	CAG	CGG	G
A	ATT (Ile/I) Isoleucine	ACT	AAT (Asn/N) Asparagine	AGT (Ser/S) Serine	T
	ATC	ACC (Thr/T) Threonine	AAC	AGC	C
	ATA	ACA	AAA (Lys/K) Lysine	AGA (Arg/R) Arginine	A
	ATG[A] (Met/M) Methionine	ACG	AAG	AGG	G
G	GTT	GCT	GAT (Asp/D) Aspartic acid	GGT	T
	GTC (Val/V) Valine	GCC (Ala/A) Alanine	GAC	GGC (Gly/G) Glycine	C
	GTA	GCA	GAA (Glu/E) Glutamic acid	GGA	A
	GTG	GCG	GAG	GGG	G

RNA code:

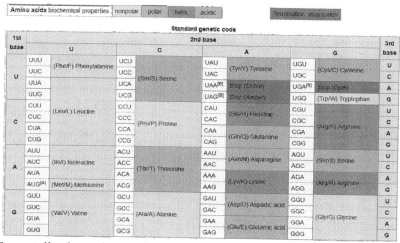

Amino acids biochemical properties: nonpolar | polar | basic | acidic   Termination, stop codon

Standard genetic code

1st base	2nd base U	2nd base C	2nd base A	2nd base G	3rd base
U	UUU (Phe/F) Phenylalanine	UCU	UAU (Tyr/Y) Tyrosine	UGU (Cys/C) Cysteine	U
	UUC	UCC (Ser/S) Serine	UAC	UGC	C
	UUA (Leu/L) Leucine	UCA	UAA[B] Stop (Ochre)	UGA[B] Stop (Opal)	A
	UUG	UCG	UAG[B] Stop (Amber)	UGG (Trp/W) Tryptophan	G
C	CUU (Leu/L) Leucine	CCU	CAU (His/H) Histidine	CGU	U
	CUC	CCC (Pro/P) Proline	CAC	CGC (Arg/R) Arginine	C
	CUA	CCA	CAA (Gln/Q) Glutamine	CGA	A
	CUG	CCG	CAG	CGG	G
A	AUU (Ile/I) Isoleucine	ACU	AAU (Asn/N) Asparagine	AGU (Ser/S) Serine	U
	AUC	ACC (Thr/T) Threonine	AAC	AGC	C
	AUA	ACA	AAA (Lys/K) Lysine	AGA (Arg/R) Arginine	A
	AUG[A] (Met/M) Methionine	ACG	AAG	AGG	G
G	GUU	GCU	GAU (Asp/D) Aspartic acid	GGU	U
	GUC (Val/V) Valine	GCC (Ala/A) Alanine	GAC	GGC (Gly/G) Glycine	C
	GUA	GCA	GAA (Glu/E) Glutamic acid	GGA	A
	GUG	GCG	GAG	GGG	G

Essentially, the codes are the same except that thymidine in DNA is replaced by uridine in RNA. In addition, RNA has two hydroxyl groups on the ribose ring. These simple changes account for all the different physical properties, including instability, that RNA has compared with DNA. Nonetheless, when it comes to coding, the purpose is identical. I show the different structures and discuss the differences a little more in Section 5.

Also, what you see in the tables from Wikipedia is organization of the amino acids into different categories: non-polar, polar, basic and acidic. These differences are what affect the functionality of protons. By ordering amino acids with different properties and sizes in different sequences, you can create structures with completely different and highly specific purposes. This was covered sufficiently in the first section.

REDUNDANCY IN THE DNA CODE

The genetic code is described as degenerate, and contains redundancy. What that means is that a single amino acid can be coded for by more than one codon, but a codon only ever specifies one amino acid. This may have advantages in terms of mutations, in that amino acids that have a higher number of codons coding for them are more tolerant to mutations. Also, there has been shown to be a difference in speed between translating codons that specify the same amino acid. The implications of this are not fully understood.

JUNK DNA

For many years, large sections of various genomes appeared not to code for anything, as they did not code for proteins. These were assumed to be evolutionary leftovers--parts of the code that were no longer required from previous iterations of the species, or indeed other species. It is looking increasingly that these ideas were completely wrong, and that huge sections of DNA that don't code for proteins and were thought to be junk play a role in the timing of which segments of DNA get copied. It would appear that there is more code within the genome than was first thought.

NON-CANONICAL CODE

In the main section of this book, I cite the frozen accident as evidence against the possibility that the DNA code (or an RNA code) could have evolved. The frozen accident expression was coined by Francis Crick and relates to the fact that the DNA code and translation machinery is universal for all organisms, and has been the same since LUCA, which appeared very shortly after the Earth was able to first support life. The code has not changed, it is frozen, and I went into the reasons why it is frozen in the main section--namely, that any significant changes or additions in the code would require too many coordinated fortuitous mutations. I used the HIV virus as an example and how unlikely it would be to find more than two beneficial mutations in one replication cycle, let alone the dozens or more that would be needed for a change to the code. However, there are examples of organisms, and structures that contain non-universal code, some of these are discussed here.

CODON REASSIGNMENTS

Codon reassignments are when a codon changes its assignment or meaning. This change can be from one amino acid to another (sense to sense); from a stop codon to amino acid (nonsense to sense), or vice versa; or from assigned to unassigned, or vice versa. In total, there are thirty-four codon reassignments that have been discovered[105]. Most of these are in the mitochondria of cells where "alterations appear to be facilitated due to their reduced genome size and complexity, which encodes only a small set of essential genes." The genome is much smaller because the mitochondrial DNA only has one chromosome and nuclear DNA has forty-six chromosomes. Also, mitochondrial DNA codes for processes that specifically take part in the mitochondria and are related to the generation of energy, so mutations are less impactful. Mitochondrial DNA translation is much more error-prone than the translation processes of the nuclear DNA, partly because replication occurs more frequently.

Fewer of these codon reassignments have been found in nuclear DNA, and the majority involve a change from a stop codon to an amino acid. (The only sense-to-sense change that has been found in nuclear DNA is in candida, a fungal species.) Most reassignments are due to changes to components of the translational machinery, e.g. ARS, or misreading tRNAs or the signaling molecules involved in recognizing stop codons.

The vast majority of species function using the universal code in their nucleus. These very minor changes, either in the mitochondria or nucleus, do not infer evolution of the code and, if anything, maybe point to the difficulty of changing the code. After all, most of the changes in nuclear code involve the reassignment of a stop codon, and therefore would have far fewer downstream implications. A sense-to-sense change would mean many alterations in many proteins. The fact there is only one known example in nuclear DNA points to the fact that, although there have probably been trillions of trillions of similar types of change, all the others were catastrophic and generated non-viable organisms. To me, this data speaks to the "frozen" nature of the frozen accident.

What is of greater interest, and perhaps less easily dismissed is the "expansion" of the code.

105   Non-Standard Genetic Codes Define New Concepts for Protein Engineering; Bezerra AR, Guimarães AR, Santos MAS. Non-Standard Genetic Codes Define New Concepts for Protein Engineering. Life. 2015; 5(4):1610-1628

EXPANSION OF THE GENETIC CODE

There are two potential examples of natural expansion of the genetic code: the incorporation of the non-canonical amino acids selenocysteine (sec) in a wide range of prokaryotes and eukaryotes, and pyrrolysine (Pyl) in the archaeal Methanosarcina genus, producing novel classes of proteins.

So what is going on here? Does this melt our frozen accident argument? Surely if the code has expanded twice, this raises two fundamental points that support random processes:

1. While changes might seem impossible, they have obviously occurred, so could have occurred many times.
2. This data shows that the code can "evolve," and therefore it could have evolved from a smaller/simpler code.

We need to look closer to understand exactly what each of these two changes infers. This is from the 2015 paper in *Life* that I cited earlier[98]:

"Incorporation of Sec in response to an in-frame UGA codon is achieved by complex recoding machinery that informs the ribosome not to stop at this position. The mechanism is distinct in prokaryotic and eukaryotic organisms, but there are some similarities. Both have a special Sec tRNA, which is a minor isoacceptor derived from a serine tRNA (Figure 2C). The other key players are SelB and SECIS (selenocysteine insertion sequence). Since Sec has its own tRNASec, biosynthesis begins with SerRS acylating tRNASec with serine, producing Ser-tRNASec. Then, different enzymes convert Ser-tRNASec into Sec-tRNASec […]"

A few important things to note from this: Firstly, both trees of life have organisms present with various important molecules central to the process of translating the modified sequence and producing proteins in the ribosome. So while sec is not part of the universal code, it was included or added within a very tight timeframe of the universal code coming into being, and before life split.

Secondly, the words "complex recoding machinery" really ring alarm bells. Firstly, sec does not have its own ARS (which would be called secRS); instead, serine is added to a modified tRNA, which is specific to sec, then changed into sec using some specialized enzymes. Then more complex machinery is deployed to ensure that a stop codon is, in fact, read as sec instead.

Both of these bits of information have implications for the interpretation of the existence of this additional piece of code. Given that they formed at around the same time as the universal code, there would not have been much time at all for them to evolve. It is very hard to see an evolutionary route from a system that just codes ser and a stop to this system that has a highly complex modification. It is possible that horizontal gene transfer could account for it. (This is the way that a lot of evolutionary theory is being forced to go now.) The other explanation is that sec was originally a part of a twenty-two amino acid universal code, but its use has been confined to a small number of organisms over the years. So instead of the code "expanding," it has in fact shrunk. However, if that were the case, wouldn't the organisms have retained a specific ARS to sec instead of needing to hijack the serine ARS? The relevance of this, if any, is unclear. This brings us on to the second example:

"While Sec is generated by a pretranslational modification of Ser-tRNASec (Figure 2D), pyrrolysine (Pyl) is directly attached to tRNAPylCUA by PylRS in response to an in-frame UAG codon in the Methanosarcina barkeri monomethylamine methyltransferase gene [12]. These are methane-producing organisms and Pyl is necessary for methane biosynthesis from methylamines."

In this instance, the Pyl code and its tRNAs and ARS are both present but, once again, it is a stop codon that is used. Again, very importantly, these are present in both archaea and bacteria, suggesting that:

"Pyl arose before the last universal common ancestral state"[106]

---

106   Structure of pyrrolysyl-tRNA synthetase, an archaeal enzyme for genetic code innovation:Kvran, J et al; Proc Natl Acad Sci U S A. 2007 Jul 3; 104(27): 11268–11273

So let's go back to the two points or questions that the idea of the code expanding could melt the frozen accident argument:

1. While changes might seem impossible, they have obviously occurred, so could have occurred many times.

2. This data shows that the code can "evolve," and therefore it could have evolved from a smaller/simpler code.

To the first point, these two examples show very rare "additions" to the code, if they are in fact additions at all, thus exemplifying the difficulties associated with the idea of "expanding" or evolving the code. It hasn't happened many times.

To the second point, and more importantly, when we look at these two examples of "expansion" more closely, we start to see they may not in fact be expansion, especially in the instance of Pyl. If anything, Pyl points to a "shrinking" of the code. If any de-evolution has occurred--namely, if organisms did start at a thermal vent, as touted by some origin of life researchers (but not Sutherland), as they moved into environments where they could acquire energy from sunlight, then the need for Pyl disappeared, the amino acid became redundant and the system simplified. If you believe in evolution and the appearance of life 3.8 billion years ago, then it would appear most likely that Pyl may have been part of an early, larger, universal code.

The Sec amino acid and machinery could also have been part of a similar larger universal code, and rather than keep all the code and machinery etc. for separately processing sec, it ditched the secRS. This seems unlikely. It does seem more likely that it was, perhaps, a later addition to the code, prior to LUCA, as it has the feeling of an afterthought. Whatever the reason and exact timeframes, it is very important that both of these amino acids and their code were present at the same time as LUCA, and were therefore most likely a part of LUCA's code. The code has not evolved or expanded since LUCA, and there is no real evidence that it did before LUCA.

The frozen accident argument remains intact; any cracks that might have appeared are really like pressure cracks on a lake that have refrozen. The only expansion of the genetic code since LUCA is that which may be occurring in labs as described in the 2015 paper but, as with the natural examples, major modifications are very hard, even using our considerable abilities and technologies. For that same reason, changes that are too big require completely new ARS, tRNAs etc., which also have to be coded for and, moreover, you need to have an understanding of where to put the new code into the sequence so that the changes you make in the end product will result in a viable protein. No mean feat, even given our abilities, and definitely beyond the abilities of random natural processes.

IS DNA A CODE?

There are some who say that DNA is not a code, and that the use of the word code is only an analogy. However, all the textbooks and scientific literature refer to it as a code. It is exactly like a code, so it is a code; but just because I say it is a code doesn't mean anything. Hubert Yockey was one of the foremost experts on bioinformatics. He was part of the Manhattan project, and was widely published in academic literature. He was not a proponent of intelligent design, in fact, on his blog[107] he was quite hostile to ID. However, he was adamant in his book, *Information Theory, Evolution, and the Origin of Life*, that when you measure the properties of DNA against our understanding of what a code is using various principles, such as Shannon's principle, then it is entirely correct to describe DNA as a code. I read this book a while back, and to go into the details of his thesis is beyond the scope of this text, but here is a quote from the book, which I pilfered from Perry Marshall's website[108]:

---

107   http://www.hubertpyockey.com/hpyblog/

108   https://evo2.org/dna-atheists/dna-code/

"Information, transcription, translation, code, redundancy, synonymous, messenger, editing, and proofreading are all appropriate terms in biology. They take their meaning from information theory (Shannon, 1948) and are not synonyms, metaphors, or analogies." (Hubert P. Yockey, Information Theory, Evolution, and the Origin of Life, Cambridge University Press, 2005)

What Yockey does not do at any point is speculate on the metaphysical implications of the code or of its origins. The furthest he will go is to say that the code came into existence and that the scientific problem of the origin of life (or code) is unknowable or unsolvable. This is becoming the default position among scientists who have a deep understanding of the science, and who are either wedded to an atheist belief system, or who don't wish to publicly be outed as "crazy creationists."

Others like Michael Behe, Stephen Meyer and Perry Marshall (the latter's expertise also lies in information theory) suggest that, since there are no known examples of a physical code appearing spontaneously or evolving due to random natural processes outside of a DNA-based system, and since there are millions of examples of codes appearing due to the efforts of intelligence, whether they be human in the form of language or computer code, or in the animal kingdom such as bird song or mating calls, it is reasonable to conclude that the DNA code provides evidence of intelligence in the origin of life.

I obviously agree, as I use this is as my one and only piece of evidence for intelligence. I believe that it is entirely logical and reasonable to do this. I do not say that it is overwhelmingly compelling evidence but, given the absence of any other viable explanations for the appearance of the DNA code, and given the significant evidence against natural random processes being able to generate the code and its translation machinery, it is entirely logical to conclude that DNA was generated by an intelligent being.

To solidify this central argument, you need to consider the alternative conclusions you could draw. We have a few facts, and other than the first, all are indisputable. The first one is only disputable if you are not prepared to accept the thoroughly researched opinion of one of the world's leading experts on information theory. Otherwise, it is a fact.

Fact 1: DNA is a code.

Fact 2: We know of millions of codes.

Fact 3: Other than DNA, every single one of those codes has a proven intelligence as its source.

Fact 4: In the entire sum of human knowledge, there are no codes that we know of that have been shown to have been generated by a random natural process.

Since these are facts, they have the potential to be used as evidence in support of an argument related to these facts. That argument can be summarized in a single sentence. The beginning of the sentence is the summary of the facts, or evidence, the end is the conclusion we could reasonably, or logically, draw from these facts. There are a number of choices for the second half:

DNA is a code and codes are only known to have been generated as the result of intelligence which...

1. proves that DNA was the result of intelligence
2. suggests that DNA was more likely than not to have been the result of intelligence
3. says nothing about the origin of DNA
4. suggests that DNA was unlikely to be the result of intelligence
5. proves that DNA was generated from purely random natural processes

Being entirely objective, conclusion 1 is a stretch, but some fundamentalist creationists would argue it is valid. I will come back to conclusions 2 and 3. To arrive at the statement concluding with 4 requires a fundamentalist's belief in naturalism. How could those facts be remotely supportive of the belief that life was not the result of intelligence? Obviously 5 is off the chart, and you might as well have said that the facts prove that, in an alternate universe, two cows simultaneously jumped over the moon, hi-fived each other as they passed, AND learned Arabic while they were travelling.

So what about 2 and 3? These are the two most reasonable statements that could come from either those who believe in an intelligent creator, or those who believe that life arose due to random processes. However, while I know that it is hard to be objective, I cannot see how anyone could reasonably argue against statement 2 being entirely consistent with the facts. Likewise, if you accept the facts, then I do not believe that statement 3 is an objective conclusion to draw; the evidence points in a clear direction, it doesn't point nowhere. You have to entirely ignore the facts presented, i.e. the evidence, to believe statement 3 to be correct.

Statement 2 best summarizes the only logical conclusion that you can objectively draw from our extensive knowledge of DNA and of codes.

SECTION 5: THE CALCULATION

The Probability of a Functional Protein Forming by Random Chance

There have been lots of calculations of the likelihood of a functional protein coming into existence through random natural processes. I even did my own back in 2008, which I published on my website, and like all the others, they have huge numbers attached to them. I will choose the one that was used by Stephen Meyer in his excellent book, *Signature in the Cell*, and which is based on better assumptions than mine. The calculation is summarized on a website you can view yourself[109] and which has a very amusing analogy applying the math of the problem to an imaginary scenario. Meyer does a much more thorough job of examining the statistics than I will here. I think I've got my point across without going into the levels of minutia that were necessary for him to go into, to fully close the loop on the argument of whether the "theory of chance" is a viable explanation of how DNA and life could have come into existence.

Firstly, there's the issue of deciding the minimum length of a functional protein. Most functional proteins in the simplest life forms, Archaea, range from 156-283 amino acids long. Shorter polypeptides (chains of amino acids) do exist, but these do not perform functions in the sense of active mechanical stuff like catalyzing reactions. Shorter polypeptides are usually just for signaling, and are mostly "passive" in their functionality. These would not be adequate to help kick-start early life, so for the calculations quoted in Meyer's book, they settled on a minimum length of 150 amino acids.

The next problem is defining a functional protein. For a protein to be functional, it needs to be able to fold into a stable, three-dimensional structure. The vast majority of random chains of amino acids do not do this. Douglas Axe, a Ph.D. Chemical Engineer, calculated[110] the proportion of chains that

---

109   http://www.originthefilm.com/mathematics.php

110   Douglas D. Axe, "Estimating the prevalence of protein sequences adopting functional enzyme folds," Journal of Molecular Biology 2004 Aug 27;341(5):1295-315

would be functional out of all the possible random 150 amino acid chains. Axe estimated that only one in 10^{74} chains of 150 amino acids would fold and be functional.

Next there is the issue of chirality. Only left-hand molecules could be incorporated for the molecule to be functional, and since any amino acids that had been produced naturally in the early Earth environment would be a racemic mix of D and L, there is a 1 in 2 chance of incorporating the right one each time. That is like flipping a coin 150 times and expecting to get heads each time, and the chances of that are roughly 1 in 10^{45}. But the misery doesn't end there.

The next issue is that amino acids can react together in more than one way. We want peptide bonds (the bonds that exist between different amino acids in proteins) to form, but other types of bonds could form that would make molecules that aren't actually proteins. Meyer estimates that the chance of this happening each time is roughly 1 in 2. This has to happen 150 times, so we have another 10^{45} to add. To get the overall probability, you add the exponents $10^{74+45+45}$ which equals 1×10^{164}, the number I give in the main section of this book. I give some examples of big numbers and relate it to the universe, but the chaps at Illustra[111] have produced an excellent video that shows the absolutely crazy nature of the number in relation to believing that a functional protein could form by random natural processes...or by the theory of chance.

---

111   http://www.originthefilm.com/mathematics.php

PROBABILISTIC RESOURCES OF THE UNIVERSE

The second number I gave is the probabilistic resources of the universe. Again, various people have attempted this, and I quote from Meyer's book:

"Dembski[112] had calculated the maximum number of events that could actually have taken place during the history of the observable universe. He did this to establish an upper boundary on the probabilistic resources that might be available to produce any event by chance. There are10^{80} elementary particles, and 10^{16} seconds since the big bang, and the smallest unit of time is the Planck time of 10^{-43} seconds. Using these numbers Dembski was able to calculate the largest number of opportunities that any material event had to occur in the observable universe since the big bang since elementary particles can interact with each other only so many times per second (at most 10^{43} times), since there are a limited number (10^{80}) of elementary particles, and since there has been a limited amount of time since the big bang (10^{16} seconds), there are a limited number of opportunities for any given event to occur in the entire history of the universe. The total number of events that could have taken place in the observable universe since the origin of the universe at 10^{139}."

According to Wikipedia, this number has now been revised to 10^{150}. Seth Lloyd[113], a professor of mechanical engineering and physics at MIT, has also performed a similar calculation using different criteria and arrived at the number of 10^{120}.

None of these comes close to 10^{164}, even if you take the highest number.

---

112   William A. Dembski (2004). The Design Revolution: Answering the Toughest Questions About Intelligent Design

113   Computational Capacity of the Universe: Lloyd Seth; Physical Review Letters, vol. 88, Issue 23, id. 237901

Obviously these numbers are just to show how silly it is to believe that proteins or other functional biomolecules could have come into existence by random natural processes since the particles of the universe are not all amino acids (in fact, none are, since it was sub-atomic particles that were considered in this calculation) and the universe hasn't spent every possible moment since it came into existence focusing on this task.

The fact is that a protein or similar functional biomolecule could not have come into existence by random natural processes.

There are no theories suggesting a stepwise approach since there are no simpler chemical structures capable of the complex functionality of proteins; functionality that would have been necessary to get life going.

SECTION 6: DNA, RNA, AND PROBLEMS WITH RIBOSE AND ANTICODONS

DIFFERENCE BETWEEN RNA AND DNA

The two differences between RNA and DNA are:

1. All RNA nucleosides have two oxygens at the bottom of the sugar. DNA is in fact "missing" one, hence the deoxy.

2. In DNA, thymine replaces uracil. This is the difference of a single methyl group on the nucleoside base.

Otherwise, the two are the same. However these two changes alter the stability of the two molecules (RNA is much less stable/more reactive) and allows RNA to fold into more complex structures.

Anticodons

The sequences of nucleosides on tRNA that align with the codon sequence on the mRNA in the translation process are called anti-codons. They are three nucleosides long, as are the codons. They are not the same, but are able to align in the same way that nucleosides align in a helix. The anticodon nucleosides align as follows: A – U, C – G, G – C, U – A. So the codon UCA (TCA in DNA) aligns with the anticodon AGU on tRNA.

PROBLEMS WITH RIBOSE CHEMISTRY AND CHIRALITY

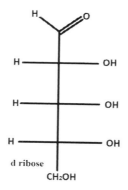

The ribose ring of any nucleoside, RNA or DNA, is derived from d-ribose (above). Ribose itself could form a number of different chiral forms, since there are three chiral centers with two possible options for each center and, therefore, nine possible forms could come into existence if random processes were generating ribose rings.

Moreover, d-ribose can form the classic pentagon ring that we see with nucleosides or the hexagon, and either of these has two possible forms:

Nucleosides have a specific base at the 1′ position on the ribose ring, and this is represented by R.

In general, the multiple forms that ribose could adopt, along with the difficulties of performing chemistry on it in the first place, add to the obstacles to creating precursor molecules via a prebiotic soup. You then need the bases, purines and pyrimidine rings. Suffice to say, it only adds to the already impossible chemical barriers encountered.

SECTION 7: RNA WORLD AND APTAMERS

The RNA world is a red herring, but it is the most commonly cited red herring in discussion on the origins of life. It has been around for a long time, in spite of the fact that it is a weak theory. The reason it has persisted is that it is the best that the naturalist establishment has in its defense of belief in random causes. So what is the theory?

RNA, as shown in Section 5 of the Appendix, has a different structure from DNA, namely a slightly different pyrimidine in the form of uracil. The presence of the extra oxygen atom on the ribose ring, and the loss of the methyl group on the pyrimidine, change the properties of chains of RNA when compared to DNA. Since they are nucleosides, and only four are used, they are just as able to represent a code as DNA. As we have seen in the main section, RNA is often used inside the cell as a mobile form of DNA, and is used as a template to make proteins. However, the changes in the structure of RNA compared to DNA mean that RNA has extra abilities.

For one, it is able to fold into multiple shapes or structures, in a similar way as proteins. As a result, chains of RNA are able to act as enzymes in the same way as proteins, albeit much less efficiently. Enzymes are proteins that catalyze reactions. They are present in just about every cellular process, and cause specific chemical reactions to happen that otherwise wouldn't. We became very familiar with a very important group of enzymes in the chapter where translation was described. The ARS proteins are a group of enzymes that catalyze the addition of an amino acid to the corresponding tRNA. As I was at great pains to point out, as it is central to the issue of the chicken and egg paradox, without these ARS proteins, the tRNA would never associate with the amino acid that it is assigned to. The ARS proteins have very specific active sites that bind the amino acid and the tRNA separately, then bring them together. Without the enzyme, this would not happen. Enzymes perform tens of thousands of different operations like this in cells billions of times a day.

They are phenomenally accurate and fast.

Ribozymes are the RNA equivalent of amino-acid-based enzymes. They are rare in the natural world compared to proteins, however thousands of different ribozymes have been synthesized in the lab, to the extent that, for many of the tasks an enzyme can perform, labs have been able to create a ribozyme that is able to perform the same function. The one big difference is that proteins are much better, since RNA is less well-equipped to perform the equivalent enzyme task. However, in principle, it is possible that RNA could once have done some or all of the jobs that are now assigned to proteins.

So RNA has two key properties. Firstly, it can and does act as a store of genetic information and, secondly, it can act like an enzyme (protein). It is therefore not a huge stretch of the imagination to envisage a world in which only RNA existed and did both tasks, thus potentially bypassing the chicken and egg question.

The sequence of events goes something like this:

1. RNA is produced via some as yet unidentified route in the prebiotic world (I will come onto pRNA and PNA etc. later).

2. Trillions of chains of RNA formed spontaneously and eventually one was capable of self-replication, the key step on the route to becoming a lifeform.

3. This self-replicating RNA molecule then acquired other abilities that further enhanced its self-replicating qualities, including producing other RNA molecules that assisted in the process.

4. Eventually, proteins came on the scene and identified themselves as much better than RNA at doing enzyme stuff.

5. RNA learned how to code for these proteins while retaining its self-replicating abilities.

6. DNA then lumbered onto the scene, or RNA spontaneously shed its extra oxygen, realizing that it wasn't very stable.

7. Et voila, we have the modern system with RNA, once king of the molecular world, now relegated to a few ceremonial roles in the production of proteins.

8. Oh yes, almost forgot…somewhere along the way a cell wall appeared.

There are different variations to this but, in general, that is the sum of what supposedly happened. The evidence that is used to support this hypothesis is the fact that you have rRNA and tRNA present at the heart of the translation process in the ribosome. These are described as biochemical fossils pointing to the RNA world. However, since these molecules do not actually do the translating, then I would argue that they are not fossils pointing to a past translation role; rather, they are present specifically for the purpose of acting as templates or to bind and hold mRNA. Anyway, I will not dwell on that, as the RNA world is rich in much bigger targets than this.

Now there are the standard objections to the RNA world, such as the issue of RNA's instability. It is indeed true that an RNA chain would not have stuck around long enough to have afternoon tea, let alone construct its own world, so various solutions have been proposed to this. These include the PNA world, which is a simpler and far more stable molecule that could have converted to RNA at a later stage. However, these objections and solutions are a distraction, as they are relatively minor compared to more significant and fundamental problems with the theory. Not only is the stability of RNA the least of the RNA world's concerns, but the solutions they propose create even more problems. I will briefly touch on these at the end of this discussion.

Then there are all the other arguments against such a scenario, some of which we are very familiar with. Getting starting materials in sufficient quantities and purities. Generating long chains of homochiral molecules spontaneously in adverse chemical conditions. The statistics issue of producing a useful biomolecule. Leslie Orgel, one of the first proponents of the RNA world, said that the chances of producing a self-replicating molecule are one in 10^{48}. Then you would have to do the same again several times over to produce other functions. Then there is the issue that, in a lab, in spite of huge efforts by people such as Harvard researcher Jack Szostak[114], they have as yet not been able to produce a fully self-replicating molecule. All of these present huge, arguably insurmountable obstacles to the viability of the RNA world in themselves. But there are two far more important issues that are related to each other and that lie at the core of the chicken and egg and coding problems.

Now, let's assume for a moment that the RNA world actually existed, and that all the arguments I just cited had been overcome (although in reality they wouldn't have been). We have this wonderful system composed of RNA, which self-replicates perfectly and has generated a number of useful ribozymes, if needed, to help speed that process up. This is our RNA world. But there is a problem, and that problem shows why this is one of the greatest red herrings ever generated. The fact is that the RNA world does not take us one step closer to the current system than saying that a bucket of water solves the problem.

Why do I say that?

When people present the RNA world as a route to life as we know it, and they briefly sum up objections, as well as listing the ones I mentioned above, they may quickly throw out a phrase like, "Of course, at some point proteins would take over and the RNA would learn how to code for them." The curious might, of course, ask "And how would that happen?" The reply might be, "We are working on that," or, "aptamers are the clue to solving that problem." At which point someone who has not availed themselves of the information on this subject to the extent that I and many others have, would nod and make a mental note to look up aptamers when they get home.

I will come on to aptamers in a bit, but first I want to explain why the RNA world brings us no closer to solving the central problems associated with believing in random processes being behind the appearance of the

---

114   https://molbio.mgh.harvard.edu/szostakweb/researchRNA.html

DNA code and its translation machinery:

1. You do not have proteins, and no one has ever proposed (sensibly) how we would get them.
2. There is no code, nor a viable route to a code.
3. There's no DNA, and only vague notions of how it might transition.

I will show that all of these are true in a moment, and why they are so problematic in terms of generating solutions, but let's first understand why these observations about the RNA world mean that it doesn't get us any closer to solving the puzzles at the heart of the origin of life.

The lack of proteins is a killer issue, and no one has ever shown in a viable A-LUCA sequence how on Earth this could happen. In fact, from my understanding of the state of research on this topic, they only have A and H...every step in between is missing. You hear phrases like "protein takeover" being thrown around like it is some sort of biochemical turf war. Or you hear "RNA sequestered existing proteins" without ever explaining how. These are all Genesis 1:3.

No one has been able to propose a viable route because it is inconceivable. There are only two possible sequences of events in which an RNA world might transfer to an RNA/protein world.

The first is that proteins appeared by themselves and RNA somehow incorporated them into the system. Firstly, we have already overcome a multiverse of statistical hurdles to create an RNA world, we now have to have another multiverse to create a useful protein. Also, the chemistry required to make proteins is completely incompatible with the chemistry that makes RNA chains, so there is no chance you'd ever find them together in a natural non-biological environment. But let's just say they could, and that we are now in a multiverse of multiverses and you have an RNA world happily spurting out copies of itself and a protein comes along that is able to give a hand, for want of a better expression. How does the RNA world "sequester" this useful protein?

Silence.

There is no way that it could conceivably do this. The self-replicating RNA molecule has become perfectly adapted at self-replicating. The sequence of nucleotides of this self replicator is specific to the ability of self-replicating, and nothing else. It has no additional opportunistic abilities of "spotting" a useful passing protein and incorporating that into the chain, as the chain is not adapted to perform that function. It is nonsense to anyone who considers this for more than half a second, and yet it is put out there out there by desperate atheist scientists trying to cover up the emperor's naughty bits. Of course, to the scientifically illiterate it works a treat, but to anyone else who is equipped to challenge them, the ludicrous nature of the idea is obvious. A protein might hang around and help for a bit, but within a couple of replication cycles, it will degrade and that fortuitous encounter would have been for nought. Ultimately, the RNA molecule is not equipped to code for a protein, so it has no means by which it can incorporate it into its world.

The second potential route that is mooted is that the RNA "invents" a protein. Some ribozyme(s) with an ability to create proteins appears. What I mean by that is that a ribozyme appears that is able to grab amino acids from the non-existent soup of amino acids and start joining them together, thereby creating polypeptides. There are huge problems with this. Firstly, once again, a ribozyme like this would have to randomly appear. The statistics against that occurring are enormous even if the chemistry was amenable to it happening. Secondly, the ribozyme would then be producing random sequences of amino acids, most of which would be useless polypeptides, not proteins, and it is even less likely that a specific protein that would help speed up the process of (self)-replicating the RNA. In the third multiverse, in a nest of multiverses that this has actually happened in, there is now a big problem facing the system. There is no code for the protein--no way of passing on this amazing discovery, so it would suffer the same fate as the random protein that appeared in the first "route", it would eventually suffer chemical degradation and be forgotten. Back to square one.

There is ultimately no conceivable route by which proteins could appear and become incorporated in an RNA world. There is no route by which a protein takeover could occur, and those who use expressions like that are being disingenuous. I can name at least two popular science writers who deploy these expressions, and they should be treated with the disdain that they deserve, but they aren't because that is not the narrative that the secular establishment wishes to support.

So that is the first reason why the RNA world is a red herring, as it never gets close to explaining the appearance of proteins...no closer than a bucket of water.

The second problem, the fact that the RNA world does not actually contain a code, is, if possible, even worse. Let's face it: a theory that is supposed to explain the appearance of a code should at least have a code, but it doesn't. Codes containing information are abstract; their meaning is not obviously related to the medium, or structure of the medium. You need a "dictionary" to decode it. There is nothing abstract or code-like about the RNA molecules in the RNA world described by its proponents. Their structure is purely functional and not informational in the way that DNA is informational. The sequence of nucleosides in a ribozyme or self-replicating RNA molecule is not a sequence with an abstract meaning, it would have evolved to be perfectly adapted to self-replicating, and every nucleoside in the sequence would have the sole function of providing structural or chemical support for the process. There is no abstract meaning in the sequence. There is no code.

There is no viable or conceivable route by which this type of sequence would become a code for proteins. Think about it for a second. If such a system had come into existence then, as stated above, by the laws of evolution, the sequence of a self-replicating molecule would be best adapted to performing the task of self-replicating. There is no need for a code, let alone one that codes for an entirely different chemical system. Any alterations in the sequence would have to further improve the ability of the molecule to self-replicate, or it would be discarded. Furthermore, any such code would have to appear "unknowingly."

The RNA molecule cannot move towards an "intended" endpoint as it is incapable of intention, and the idea that it could invent a code is inconceivable.

The third problem--namely, the problem of transitioning to a DNA system--should be easy to solve, but in reality, it is not. This, again, comes down to the point that a hybrid system would not work. Firstly, we wouldn't have an RNA system coding for proteins for the reasons outlined in the previous paragraph. But even if you did, the translation machinery would be adapted to read RNA molecules. While DNA molecules are very similar, any proteins would have highly specific active sites that might not be able to fit the DNA molecules. More importantly, where would these DNA molecules come from? Why would they appear? How would the system know how to make them? It's fantasy.

APTAMERS AND ESCAPED TRIPLETS: DOWN THE RABBIT HOLE

Below is an excerpt from a series of articles on the BioLogos website. These articles were mainly focused on refuting the arguments in Stephen Meyer's book, *Signature in the Cell*. As I mentioned before, this book is excellent, and I cannot fault the conclusions drawn, even if it took a very long time to get there. Anyway, this is what Dennis Venema said in 2016:

> *"So, is there evidence that amino acids can bind directly to their codons or anticodons on mRNA? Meyer's claim notwithstanding, yes—very much so! Several amino acids do in fact directly bind to their codon (or in some cases, their anticodon), and the evidence for this has been known since the late 1980s in some cases. Our current understanding is that this applies only to a subset of the 20 amino acids found in present-day proteins..."*

And again:

"The fact that several amino acids do in fact bind their codons or anticodons is strong evidence that at least part of the code was formed through chemical interactions— and, contra Meyer, is not an arbitrary code. The code we have—or at least for those amino acids for which direct binding was possible—was indeed a chemically favoured code"

When I read this, I suddenly had one of those moments where you panic and think, "Did I miss something really obvious?" Of course, one of the central issues around the chicken and egg paradox is that the codons or anticodons do not have any direct, specific, physical association or attraction to the amino acids they code for. The codons do not specifically physically identify or select the amino acid they code for. I hope I make this point very clearly in the main text. If I'd missed something that contradicted mine, Meyer's and many others' assertion that the code is in fact arbitrary, or abstract, then I really am incompetent. Hence my panic. Then I looked at the references that he was citing as evidence and breathed a sigh of relief. I am very familiar with the work of Michael Yarus and knew instantly that Venema of BioLogos was either being deliberately misleading or had not understood the findings that Yarus had made, because not even Yarus claims that "several amino acids do in fact directly bind their codons or anticodons."

Michael Yarus first made a discovery in 1988[115] that in the binding sites of aptamers that bound amino acids, the codons that code for those amino acids were present in a higher proportion than in non-binding site areas of the aptamers. This, he suggested in *Nature*, was evidence that there is a link between the structures/stereochemistry of the codons and the amino acids they code for. Is he right?

Firstly, what are aptamers? Aptamers are man-made RNA oligonucleotides specifically designed with the purpose of binding specific chemical entities.

---

115   Yarus M., and Christian, E.L. (1989) Genetic code origins. Nature 342(6248):349-350

They are basically high specificity organic chelating agents. They have acquired significant commercial use over the years, but they have also offered hope to the beleaguered RNA world community, mostly because of Yarus' work.

So these aptamers, which are made entirely of RNA and may be 30-100 nucleotides long, fold into structures that are capable of specifically binding certain amino acids. However, on analyzing the sequences of amino acids in the binding site of these aptamers, Yarus suggests that there is an overrepresentation of codons or anticodons that code for the amino acid the aptamer is binding. In the subsequent years, during which time millions of aptamers have been made, this has been found to be true for nearly half the amino acids that are coded for in the DNA-protein code. What I mean by that is that, out of the thousands of aptamers that have been produced that bind amino acids, this association between codons and amino acids they code for has been found for half the natural amino acids. Crucially, though, and where poor old Venema is getting in a muddle, or being deliberately misleading using the language he uses, is that there are other nucleosides and codons present in the binding sites as well. It is just that Yarus' findings suggest that the codons specific to the amino acids are overrepresented in binding sites compared to areas of RNA outside the binding sites. However, without these other nucleosides, the site would not bind the amino acids and therefore, while the codon portions of the site may be involved in binding, they are not **directly** binding in the way that Venema suggests.

Let's say you have an aptamer that is sixty nucleotides long, and that there are, therefore, a total of twenty possible codons, and that the binding site is constructed using twenty nucleotides. You might have two triplets coding for the amino acid it codes for in the binding site. There are still fourteen other nucleotides in the binding site and all could still be equally involved in the binding of the amino acid, but the presence of these two codons would, correctly, be seen as an overrepresentation of that codon from a statistical perspective. This is very different from what Venema and BioLogos are inferring. They appear to be interpreting Yarus' work to suggest that the codons alone are able to bind the amino acid they code for, whereas in fact Yarus is saying that, in a site that needs, say, twenty nucleotides to selectively bind an amino acid, there is a higher chance of codons that code for those amino acids in that sequence than random statistics would predict. So what are the implications of this?

Firstly, there has been considerable criticism of Yarus' methods over the years, and his results are viewed as "controversial" in mainstream science. There is potential for selection bias with regard to the aptamers used for the analysis, and some scientists question the statistical methods that have been deployed. I have seen results in clinical papers that show no association mangled by statisticians to show strong association by one technique and none by a different technique. In all truth, I am not in a position to comment on the validity of Yarus' findings, and I don't ever really intend to be, as I believe they are irrelevant. I am going to work on the assumption that his findings are true. Namely, there is a strong association between codon sequences and aptamer binding sites. This is true for man-made aptamers and a couple of natural examples.

There are a number of points that should be noted about Yarus' theories.

1. There is no viable explanation of how the RNA world got to the state where it started using amino acids or proteins or generated aptamers.
2. There is no code. We still don't have a code. Aptamers are not a code; they are RNA molecules that bind amino acids.
3. There is still no sign of DNA, or a viable root of how to get to DNA.

So we have the same problems that I outline for the wider RNA world but, given how BioLogos make so much of this theory, I feel it is worth rebuking it as best I can. Plenty of others have done so in different ways, but here is my take on it.

To the first point. There is no answer to the significant, insurmountable challenges I suggested in my earlier discussion of the RNA world. Much of

Yarus' theoretical work[116],[117] focuses on how you would get from a system that he proposes to the current code. This is the "escaped" triplet theory. I will come back to that in a minute but, for now, using the A-LUCA analogy, he's decided to start at D and forget all about A. He provides no viable theory as to how or why amino acids and proteins would suddenly become available, and why a perfectly adapted self-replicating molecule would start using amino acids and building proteins. This is the central issue that lies at the heart of the origins of life question. Why? Even if amino acids were available, why would an RNA system sequester them?

To the aptamers themselves. These are man-made RNA molecules created in a lab using state-of-the-art technology and pure starting materials. For nature to produce one fifty nucleoside oligomer alone is something that has never been observed outside a cell, and no viable mode of production in natural conditions has been proposed. However, there would have to be millions upon millions produced for one useful aptamer to appear. It would then have to find the self-replicating molecule among the millions of useless RNA molecules, and start working with that…on what? Back to the "why make a protein?" issue. So what if an aptamer can grab a specific amino acid? The starting point of Yarus' system is where the RNA world is already building proteins using aptamers and other ribozymes, but at no point has the central issue of why it would build proteins in the first place ever been considered. That is because there is no reason why, without the system knowing it's a good idea…which of course it doesn't. (At the end of this section, I take this further.)

The problems are further compounded by the question, "What kind of proteins would this system make?" Given that there is no code at this stage, the polypeptides produced would all be random sequences of amino acids. The

116   Yarus, M., Widmann, J.J. & Knight, R. J Mol Evol (2009) 69: 406. https://doi.org/10.1007/s00239-009-9270-1

117   The Genetic Code and RNA-Amino Acid Affinities; Life (Basel). 2017 Jun; 7(2): 13. Michael Yarus

vast majority of these sequences would not even be functional, let alone have a function that is relevant to improving the system's efficiency. Such a system has no survival advantage, and no way of passing on a survival advantage, as there is no code to record and replicate improvements. Moreover, the system would rapidly disassociate and break up, as it is has no cell wall. If the system was in an enclosed space or low temperature environment with little movement, there is no way any of these molecules would be able to associate in the first place, as they would be surrounded by the useless by-products of the random processes that generated them in the first place, and unable to move away from their dim-witted cousins.

So even in the event that Yarus' observation that some man-made RNA oligomers have a higher representation of codons in their amino acid binding sites is correct, it is meaningless without showing the utility or logical origin of such molecules in an RNA world system, and he doesn't.

Then what about step D to LUCA? This is the escaped triplet theory, whereby these codons "escape" their assumed stereochemical function in an aptamer-like molecule and become codons within the code. There are a lot of scientific Genesis 1:3 "let there be light" type statements posturing as theories as to how you achieve this, like this one:

"Cognate RNA triplets within amino acid binding sites evolve to act as anticodons in tRNAs and codons in mRNAs"[118]

Instead of "let there be light," it is "let them evolve." These statements are nonsense, as they in no way address the question of how they evolve from being a 24mer, which is able to bind an amino acid to a triplet...which is, after all, the key problem. His theory explains nothing on this at all. You need 20-24 amino acids to form a binding site, and yet this is somehow supposed to "magically" evolve or escape into a triplet coding for an amino acid. As far as an explanation of how a triplet code formed, Yarus' work (which is the

118 The Genetic Code and RNA-Amino Acid Affinities; Life (Basel). 2017 Jun; 7(2): 13. Michael Yarus

best) provides absolutely no explanation at all.

Moreover, neither he nor anyone else ever explains how a small group of molecules that may have some functional utility as part of a large group of molecules acquires informational meaning all by itself. In other words, there is never a viable explanation of how the codons, which on their own have no stereochemical affinity for their respective amino acids, in spite of the apparent BioLogos inference they do, acquire the ability to code specifically for a single amino acid all by themselves.

In fact, this gets right to the heart of this rabbit hole. There is no direct stereochemical affinity between a codon and an amino acid, period, and the way that Yarus has attempted to conjure one up does not mask the fact that there is no link. His theory creates a rabbit hole without an exit or entrance and in which you could conceivably go bonkers trying to get out. Trying to make sense out of nonsense is not good for your mental health. Best to leave him and his devotees, Venema being one of them, on their own to go 'round and 'round in circles for eternity, creating new tunnels which they believe will lead them to light, but never can. They tried to solve the problem of getting from A to LUCA by starting at D, when you really need to either start at A and work forward or start at LUCA and work backwards. When you really look at Yarus' works, and indeed the whole RNA world theory, they haven't started at D at all, but at 172.

The code is arbitrary and abstract...period. The physical properties of the code have no physical relationship to the individual amino acid they code for, and there is an additional level of abstraction in that the code has no physical relationship to the structure and function of the sequence of amino acids for which they code. In other words, the properties and purpose of the proteins could not be determined from the sequence of nucleotides except by the translation of the code. The RNA world and the PNA world (an attempt to solve the chemical stability issues of the RNA world) are red herrings, and Yarus' work is a rabbit role inside the red herring. None of them explain how or why proteins appear. None of them explain viably how the code formed from these aptamers, and none of them show a fully-formed route to a DNA based system. Why anyone entertains any of these as attempts to describe a DNA-based system that codes for proteins is beyond me. For once, I agree that it is all spaghetti in the head.

The truth is that this whole distraction is a result of the blind devotion to the modern scientific establishment's mantra of insisting there must always be a natural scientific explanation. The code of today has no stereochemical basis at all. It is quite simply a fact that the individual codons are not attracted to the individual amino acids, but the implications of this are unacceptable since the code is, therefore, by default, arbitrary and, thus, could not be the result of a random natural process. Carl Woese, the famous microbiologist and biophysicist who first characterized Archaea, pointed out the lack of stereochemical basis as troubling, and so to avoid the trouble, Yarus and his devotees have tried to generate an historic stereochemical relationship where none exists now. Their error is betrayed by the fact that there is no viable route in to the system they propose from a precursor system, and no viable route out of their system to the current system.

I have described this as a rabbit hole within a red herring, but there is another analogy that keeps coming to mind. The implications that Yarus derives from his observations are like a child seeing a cloud in the sky that is shaped liked Britain and saying that therefore the cloud must come from Britain. Now if you are standing on American soil, then it is easy to explain to the child why, even though the cloud is shaped that way, it doesn't come from Britain; it is a fluke. However, by luck we are standing in London, England looking at this cloud, so it is very hard to reason with the child that, while indeed the cloud did come from Britain, it is not that shape because it came from Britain.

THE ULTIMATE HEAD-SCRATCHING HEART OF THE PROBLEM

In various previous sections, parts of the central problem have been discussed. I want to focus on trying to articulate a specific conceptual difficulty separately here, rather than get too deep into it in the main text.

In the numbers section, we learned that, whatever route you take, if you cannot conceptualize a way that a functional protein might be "developed," then it would have to form via an entirely random process with the incumbent, insurmountable statistical problems associated with this route. By default, for proteins to form *de novo* in a sequence that contains relevant function requires either foreknowledge that the sequence will have that relevant function, or that it formed randomly by surmounting insurmountable odds.

In the chicken and egg section, I show that there is no physical or chemical link between the codons, anticodons or tRNA molecules to the amino acids they code for, which makes the code arbitrary. I also show that the RNA world does not account for the formation of the code. In fact, it accounts for nothing other than an imaginary world in which RNA is a self-replicating system. I discuss Yarus' ideas and research and show once again that it does not actually account for anything, and that his central thesis is irrelevant since no viable route to DNA codon assignments is provided. In that section, I touch on what I am about to discuss here.

DNA codes for two things: the individual codons within DNA code for specific amino acids, and then specific larger sections of DNA code for function. These are the two levels of abstraction that I refer to a number of times, and they make a materialistic solution completely inconceivable by a sequence of random processes, no matter what system we invent. This is evidenced by the fact that no one has been able to conceive it. Let me clarify precisely why it is inconceivable by random processes.

Firstly, let's look at the codon assignments. Once again, no one shows in a viable manner how you would get the triplet assignments of nucleosides to amino acids. Yarus' "theory" does not show a plausible route from aptamers to triplet codons, and no one else really tries. It is pure Genesis 1:3. This is the first level of abstraction--that chemicals of one type code for a completely different type.

Conceptually speaking, there is no reason why this relationship would develop within an evolutionary context. First of all, there are no viable stepwise routes proposed, but more importantly there are no stepwise routes by which a nucleoside system would benefit at each of those steps. WHY would acquiring the ability to code for amino acids be of benefit unless the system knows that amino acids are able to come together in a way that creates function? This leads us to the even larger head scratcher if you are trying to generate a random processes route.

The second level of abstraction is even more profound: the fact that DNA actually codes for function in a separate, unrelated chemical system. For a nucleoside to learn how to code for individual amino acids and, at that same time, layer function above is the deepest problem of all. Individual codons code for specific amino acids, and yet different, unrelated, sequences of codons (or amino acids) code for function. The code for codons is actually only a code within a code. A sequence of ten codons in one gene might have a completely different purpose as the same sequence in another gene. Yes, the amino acid sequence would be the same, but the purpose or function might be the same or different, depending on the context of the surrounding sequences. Critically, the system has no foreknowledge whatsoever of what the combination of amino acid sequences will yield.

Moreover, a mechanism, if a viable one had been proposed, that would account for the nucleoside codon to amino acid relationship, would most likely be at odds with one that accounted for the longer nucleoside sequences to function relationship. Therefore, firstly, there is no benefit to acquiring the ability to code for amino acids alone. Secondly, without being able to code for amino acids alone, you can't code for function. So imagine we live in an alternate universe and Yarus was right, and that RNA acquired the ability to code for individual amino acids. The interactions that would lead to this repeatable sequence of nucleosides coding for specific amino acids would be different from the interactions that would have defined a much longer sequence of nucleosides that performed a specific function. This is because of what I mentioned before; one sequence of amino acids might have one function in one protein and a different function in another because of the different shapes that they may acquire due to the contextual sequences of amino acids on either side. This would produce different sequences of nucleosides if nucleosides were learning to code for function, because the interactions would be different due to the different shapes. The granularity of specific repeatable codon to amino acid assignments would not be retained by a sequence that is hundreds of nucleosides long trying to "capture," through chemical interactions, the function of a sequence of amino acids. And yet the two levels of coding must be present for the system to work.

This aspect of the DNA code dictates that foreknowledge of the synergistic properties of the sequence of amino acids (its function) was needed for these sequences to appear in nucleoside code. Foreknowledge requires intelligence. I do not add this to the table, although it is arguably very strong evidence requiring intelligence. It has a relationship to ID, since design requires the manipulation of things for a purpose, and that purpose is the foreknowledge.

SECTION 8: EVIDENCE OF FAKE NEWS

This is a screenshot of the results of a google search for "origin of life" on September 22nd 2018. This has been the top link on Google for a very long time. The title suggests that researchers are close to solving the Origin Of Life Conundrum, however, as I explained in the main text, the main researcher they cite, John Sutherland, himself concedes that we aren't even yet at the end of the beginning when it comes to providing explanations for the origin of life, let alone nearly solving it.

How do you feel about being lied to all these years? How do you feel about the way the establishment tramples all over the truth? How do you feel now you know the truth?

I hope you feel freedom at being released from the shackles of ignorance imposed on you by those who believe they know better. I hope you feel empowered. I hope you want to research the subject more and I hope you will share what I have shown to be the truth on this subject – believing that DNA was the result of intelligence may not be so dumb after all.

ABOUT THE AUTHOR

Dr. Orson Wedgwood attained a Bachelors in Chemistry and a Ph.D. in Organic Medicinal Chemistry in the UK. In his subsequent career he has worked alongside international experts to develop, implement, and publish scientific and clinical research on cutting-edge life-saving medicines. He is currently working for a Biotech investigating potential treatments for genetic obesity.

Orson previously published a novel called Deadly Medicine, a thriller that enjoyed sufficient success to encourage him to write more. He also hosts a popular blog and discussion forum on the science of consciousness and Near Death Experiences.

Orson is happily married to Kirsty, a novel writer from New Zealand. They recently returned to England after a seven year stint in North America.

17322641R00162

Printed in Great Britain
by Amazon